JOB HUNTING AFTER 50

JOB HUNTING AFTER 50
STRATEGIES FOR SUCCESS

Samuel N. Ray

John Wiley & Sons, Inc.
New York • Chichester • Brisbane • Toronto • Singapore

Library of Congress Cataloging-in-Publication Data

Ray, Samuel N., 1926–
    Job hunting after 50 : strategies for success / by Samuel N. Ray.
        p.   cm.
    ISBN: 0-471-53343-2 (cloth). — ISBN 0-471-53344-0 (paper)
    1. Job hunting. 2. Aged—Employment. 3. Middle aged persons—Employment. 4. Age and employment. I. Title. II. Title: Job hunting after fifty.
HF5382.7.R39   1991
650.14′084′4—dc20   90-21698

Printed in the United States of America
91 92 10 9 8 7 6 5 4 3 2 1

Dedicated to:

Charles Pletcher, founder and chairman of the board, The Transition Team. An outplacement pioneer. The person who gave me the opportunity to have a second, meaningful career.

My mom, Pauline Ray. She taught me, by her example, that helping others is the most satisfying thing in life.

My wife, Mary, and "the kids," for their moral support and confidence.

Acknowledgments

My thanks to my collaborator, Louann Werksma, who took my outline, ideas, and stacks of research material—and worked with me to write a book that, I hope, will help many find fulfillment after fifty.

Contents

Introduction

This book is for everyone who is 50, has been 50, or will be 50. It's for everyone who works, has worked, or wants to work, for monetary compensation, at a job.

The reader who is 50 or older today and is looking for a new job is at the cutting edge of several national trends:

1. A trend in which people at an age some consider "pre-retirement" find themselves unemployed after a lifetime of service, but not financially, physically, or mentally ready for retirement.

2. A trend in which many who took early retirement (either voluntarily or not-so-voluntarily) discover that retirement isn't what they expected. George Bernard Shaw said, "A perpetual holiday is a good working definition of hell." Ernest Hemingway called retirement "the ugliest word in the English language." And Pablo Casals, still giving virtuoso cello performances at the age of 80, said "To retire is the beginning of death." So, for reasons sometimes financial, sometimes psychological, a generation of "early retirees" is re-entering the work force and making job placement for those over 50 a growth industry.

3. A trend to increased numbers of "older" people in the population and smaller percentages in the traditional "entry level" age and ability range, now causing labor shortages in many industries.

4. A trend I call the "Is that all there is?" syndrome, in which increasing numbers of people who have worked

for 25 to 30 years and more in one occupation begin looking for greater life fulfillment through their work. They are saying "so long" to their "old" jobs and embarking on exciting new ventures and adventures.

This book will reveal these and other brave new worlds and give step-by-step advice for conquering them.

As for the reader who is still in what traditionally has been deemed "career prime time" (age 30 to 49), I hope this book will open your eyes and show you why the career path you mapped when you graduated from high school or college (perhaps modeled after the traditional route your parents traveled, such as continuing promotions to upper management and a comfortable retirement in your late fifties or early sixties) needs revision. Your map is outdated. New roads (and inroads) have been cut into the world of work in the past 20 years. Mergers, takeovers, leveraged buyouts, and the big "D"—downsizing—have both scarred the corporate landscape and created dazzling new vistas.

Job security in the new world of work, if not actually nonexistent, is nothing you can bank on. While it may be hard to accept that your job is not guaranteed no matter how brilliantly you perform it, and that those pension benefits you read about in your employee manual may never materialize, once you do accept and plan your career accordingly you may discover a whole new sense of satisfaction. It is exciting to be self-reliant, always alert for new opportunities. Complacency can dull the edges of your sensibilities, whereas survival requires you to hone your senses and take more risks. You may find that the life of a fit and canny jungle stalker beats that of a house pet. And when work is that exciting, maybe you will never want to retire.

This book is based on my 40 years of experience in the evolving, late twentieth-century world of work. I began at mid-century, in 1950, as a teacher in the Philadelphia school system. I left my educational career and spent 35 years in human resources management in three different companies. I changed jobs twice after the age of 50, attained the level of vice president of human resources, and experienced the thrill of re-entering the job market when a bankruptcy dissolved my employer's operations.

Now, at age 64, I'm president of The Transition Team, an outplacement firm headquartered in metropolitan Detroit that serves employers and their outplaced employees coast to coast. Our corporate clients include *Fortune 500* companies, many smaller companies, and not-for-profit organizations.

I have helped out-of-work food processors, coal miners, auto workers, machinists, salespeople, secretaries, engineers, art directors, corporate vice-presidents, and attorneys—among others—find jobs. While they have been of different ages when they lost their jobs, many have been 50 and older. All were successful, with The Transition Team's help, in finding new and rewarding employment.

In this book, I share my knowledge and experience, along with The Transition Team's effective methods, to help you custom design your career map and successfully steer your course to a full and rich work experience after 50.

We begin, in Chapter 1, with a brief look at the world of work, as it is and as it's predicted to be.

PART I | The Preliminaries

1 | Increased Demand for People over 50

Charles H.'s career is a perfect illustration of beginning a new career in the third quarter of life. When he retired in 1980, at age 55, from a Big Three automaker, Charles was a seasoned human resources professional with a Ph.D.

Most retired auto company executives who had been through the stressful 1970s in Detroit headed for the golf course—not Charles. He joined the faculty of a major university and eventually became a department head. Ten years later, at 65, he's retiring again—from the university but not from his career. He continues to work as a part-time consultant.

Charles exemplifies the emerging "young older" professional who retires from one career and begins a new one. As a result, his employers and clients get the benefit of his seasoned intelligence, in-depth experience, and faultless judgment, while he stays young, vital, and productive well into his eighth decade.

This chapter looks at the forces that are changing our economy and creating opportunities for work that is financially and personally satisfying. Where does the worker "over 50" fit into today's economic picture?

3

AN OLDER AMERICA

In recent years, the media have been featuring studies on "the graying of America," which document the fact that America is getting older and predict what effect this phenomenon will have on our private and public lives. Most such reports begin with some reference to the "baby boom," and what impact this period of unprecedented fertility has had on American life, as did the following excerpt from a report published in *Harvard Business Review* in 1978:

> As a consequence of the 43 million babies born in the years immediately following World War II, a middle-age bulge is forming and eventually the 35-to 45-year-old age group will increase by 80 percent. By the year 2030, this group will be crossing the . . . bridge to 65, increasing the relative size of that population from 12 percent of all Americans to 17 percent, a jump from 31 million to 52 million people."[1]

Some say that America in 2030 will be populated by the largest percentage of older people in recorded history. Following, in brief, is the rest of the story:

▮▮ 60 million people are 50 years and older in the United States. They comprise one-fourth of the total population but control 55 percent of the country's income and 70 percent of the assets. One half of this group— or 30 million people—are in the 50 to 64 range.

▮▮ By 2020, those over 50 will account for as much as 40 percent of the population. Between now and 2050, the population is expected to increase by one-third, but the over-55 population will more than double.

▮▮ American society has been aging for nearly two centuries. In 1800, the median age of the U.S. population was just 16. That number increased gradually over the past century and a half, but in recent years it has picked up speed. In 1986, the median age was 31. By 2000, it will be 36 and, by 2050, 42.

▮▮ After the baby boom came the "baby bust" or "baby dearth" years of 1964 to 1976. In 1957, the baby boom reached a peak of 3.7 births per family. By 1976, that number dropped to only 1.7 births per family, a figure below the replacement rate of the population. Before 2000, there could be extreme labor shortages in entry level jobs as a result. In certain areas of the country and in certain occupations, this shortage has already begun, driving up pay rates and, consequently, prices.

‖ In 1900, life expectancy for all Americans was 47. Today, it is 74 for men, 78 for women, and increasing rapidly. Some experts see a life expectancy at birth of 100-plus years by early in the next century.

Those are formidable numbers. Is it possible that, someday, it will be fashionable to be 50? When 50 is life's halfway point, will someone age 49 be considered a "youngster?"

That's very likely, since society takes on the character of its most prevalent citizens. In some ways, America has begun to recognize these population trends and adjust for them. For example:

‖ In 1987, Congress eliminated the mandatory retirement age of 70. Increasingly, employment practices and pensions that encourage Americans to retire early are being criticized as unrealistic and out of date.

‖ Instead of offering workers that dubious "golden handshake," some progressive companies, as you'll see in Chapter 2, are building "golden bridges" to encourage older, experienced workers to stay on the job longer.

‖ Colleges and universities, faced with declining enrollments in the "traditional" student population as a result of the baby bust, are successfully recruiting older students. College Board figures show that half of all students enrolled in 1990 were over 25. This points to a dawning realization that education enriches life at all stages.

Do all these statistics add up to the death of age discrimination? Are workers over 50 now in greater demand?

Not yet. Through the 1990s at least, social and corporate trends are forcing the average age at retirement down, not up.

Some sources, including the U.S. Bureau of Labor Statistics, estimate that as many as 1.5 million middle management jobs (at salaries of $40,000 and up) fell victim to the downsizing and reorganizing ax in the past decade. Many who lost their jobs were older workers who took early retirement while lower paid, younger workers took over their responsibilities.

The ax continues to fall. Following are just a few examples of the dozens of notices found in American newspapers over a recent 13-month period.

May 15, 1989   General Electric Co. to eliminate 1,400 jobs and close two factories to consolidate electric range manufacturing operations in its appliance division.

June 17, 1989   Xerox says 680 employees nationwide have agreed to leave the company voluntarily in exchange for compensation.

August 19, 1989   Cummins Engine to lay off 100 production workers in Southern Indiana.

August 26, 1989   General Electric Co. to eliminate 500 factory jobs in Louisville, Kentucky due to sharp decline in appliance sales.

October 3, 1989   International Business Machines to offer special payments to people to quit or retire at four locations, at step that is expected to cut 600 to 1,000 jobs.

November 2, 1989   Domino's Pizza will lay off nearly 100 employees.

December 7, 1989   Dollar Dry Dock Savings Bank to lay off 24 people in residential real estate lending department.

January 6, 1990   Federated Department Stores cuts 52 people from Cincinnati headquarters.

February 6, 1990   Philadelphia paper to close its Philadelphia area plant, affecting 100 employees.

February 22, 1990   Apple Computer, Inc. will reduce its work force by 400 people, or about 3 percent, as part of an effort to improve profitability.

April 19, 1990   Holiday Inns to eliminate 600 jobs.

July 17, 1990   McDonnell Douglas announced staff cuts of 17,000 from its St. Louis and Los Angeles facilities.

The huge debt burdens assumed during the merger mania of the 1980s, defense spending cuts in the wake of the fall of the iron curtain, and the cost of reducing the federal deficit are but three of the features of a changing economic picture that is leading to further job cuts.

Some trend watchers predict that the large group of baby boom workers, who are now in their 30s and 40s, will soon be pressuring for senior management slots. The force of this pressure will result in even more older workers eased into early retirement in the 1990s.

In spite of the Age Discrimination in Employment Act (discussed in Chapter 8), which was enacted in 1967, age bias still exists in the American work place and is partly responsible for the older worker being among the first considered when cuts are made.

But, experts predict that the result of earlier retirement combined with longer life will strain the seams of beleaguered government-funded health and pension programs. Then those younger workers, the same ones who wanted the oldsters to go, will be crying, "We can't pay!"

Just what is in store for us in our golden years?

RETIREMENT TRENDS

Today, when life expectancy has increased to almost 75, the average retirement age is 62—and dropping.

Historically, Americans entered the work force at an earlier age than they do today, and left it at an older age. (Many went directly from the office or factory to the graveyard.) Consider:

▮ In 1930, two of every three men 60 and over were in the work force.

▮ In 1950, half worked, and by 1980 the figure was only one in three.

▮ In spite of the declining population of younger workers, the U.S. Bureau of Labor Statistics estimates that only one in four men 60 and over will be working in the year 2000, and an even lower percentage of women.

With most Americans leaving the work force before they turn 63—and earlier—retirement becomes a longer period of life for everyone. Does that mean blissful years of self-regulated leisure and recreation pursuits? Golf and gardening and cruises twice a year?

No. That picture is pure fantasy, say many. The reality is that neither the individual retiree, nor society at large, can support the expense of so many years of retirement. Social security and health care programs are all predicted to crumble with the weight of such statistics, if they continue.

Private pension programs are at risk, too. Many retirees have already experienced the loss of "guaranteed" pension benefits. As a result, the U.S. government formed the Pension Benefit Guaranty Corporation (PBGC) in 1974. The PBGC has taken over 1,345 pension plans of companies that have sought protection under bankruptcy laws.

But don't get the impression that you're safe with Uncle Sam. The PBGC was threatened with collapse in the late 1980s because of the large number of bankruptcy filings. As with federally insured bank and savings-and-loan deposits, the security offered is limited. The PBGC is set up to pay only up to $1,857.95 per month, no matter what your benefits are supposed to be, and it doesn't insure monthly supplements that were offered to induce early retirements.

Furthermore, the PBGC can work only if the rest of the country does. Because it insures private plans covering about one of every three workers in the United States, many of which are sorely underfunded, a PBGC failure would jeopardize the retirement income of 40 million Americans. The U.S. Department of Labor, which oversees PBGC, estimates that nearly 10,000 of the 110,000 plans insured by the agency are underfunded to the tune of $45 billion.

Then there's the myth of the retirement lifestyle. Some people, accustomed to the hustle and bustle of the workaday world, regular appointments and social contacts, and a system that rewards their efforts with pay and approval, often find retirement not what it's cracked up to be.

A 1974 Harris Poll survey of retirees found that the majority of those surveyed would still work if they had not been forced out of their jobs. A 1981 poll found that 37 percent of pensioners surveyed retired unwillingly because of a mandatory retirement age, health, or disability rule. While 63 percent retired willingly, 7 percent of those later regretted the move. Nearly half of all who are retired say they don't particularly like it.

Financial constraints can also tarnish the gleam of the golden years. All those longed-for leisure activities are going up in price, to say nothing of the escalating cost of life's necessities.

So, for reasons of lifestyle and financial necessity, people are coming "out of retirement" in large numbers. Some are even going back to work as contract labor for previous employers. Many companies (see Chapter 2) have programs that allow retirees to work at assignments similar to those they left, with more flexible schedules, and still maintain pension benefits.

Others find retirement from one career an opportune time to start another. In smaller, but growing, numbers, retirees are starting their own businesses, working for nonprofit organizations, and making money at former hobbies.

Yet, in spite of this encouraging trend, it is estimated that 18 million people over 50 who want to work are underemployed or unemployed. Outdated attitudes about older workers' productivity and flexibility contribute to this statistic, but the changing character of the American work force has something to do with it as well.

WHERE THE JOBS ARE

In 1950, only about 17 percent of all people worked in information jobs. Now, more than 60 percent of the work force work with information as programmers, teachers, clerks, secretaries, accountants, stockbrokers, managers, insurance people, bureaucrats, lawyers, bankers, and technicians. By 1995, 75 percent of all jobs in the United States will be in the service sector. These jobs are in communications, finance, trade, government, real estate, and transportation. This is the "knowledge" work force of today, where intellectual acumen displaces physical strength.

Americans are equipping themselves with more knowledge than before. By 1980, one in four workers had a college degree. Higher education creates higher aspirations for career success and material wealth, however, and many of today's college graduates are disappointed to find that changes in the work place in the past two decades have led to decreased opportunity for advancement in a narrower occupational hierarchy. Layers of management have been eliminated. Today, clerical, sales, service, and operative workers are in demand, but not at salaries that would tickle the fancy of a new MBA.

The employers have changed, too. The "blue chip" names —IBM, General Motors, General Electric, and the other 997 companies that make up the Fortune 1000 list—are not today's best bet for employment. In the 1970s, 19 million new jobs were created in the United States, more than ever before in our history. 89 percent were in the service sector.

Even more interesting, 6 million of those jobs—almost one third—were in companies that had been in existence for four years and less. And they are small companies. (The Small Business Administration defines a small business as one with fewer than 500 employees.) This segment of our economy accounts for 38 percent of our gross national product, 42 percent of all sales, and 48 percent of private employment.

An offshoot of the trend to smaller, service companies is a growth in freelance and contract labor. To tighten their belts, eliminate down time, and cut the cost of payroll administration, many companies in the service sector use the services of freelance and contract workers. As a result of that trend, self-employment is a growing sector of our economy. Many freelance workers have started their own small service firms and are employing others.

It is no wonder then that Marshall Loeb, managing editor of *Fortune* magazine, advises downsized managers to head for smaller companies. That's where 83 percent of executives discharged from big companies landed in the past decade. The Transition Team's *Professional Guide to Career Life Planning* explains why "industrial life cycle" is the single most significant factor affecting one's career growth (Figure 1–1).

Figure 1-1 Industrial Life Cycle

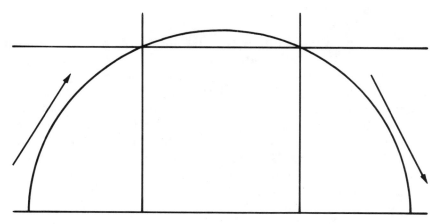

Industries generally follow a pattern of growth, beginning with rapid acceleration, leading to a period of leveling off and eventually a period of temporary or even permanent decline. . . . Industries are spawned from necessity—as the need increases for their product or service, they experience growth; as the need levels off, they do likewise. Sooner or later, the need diminishes . . .

During periods of growth, industries and their respective companies are busy hiring, training, developing, and promoting people. It is dangerously commonplace to exceed one's potential and develop a false sense of security during this time. It is also easier to shelve nonproducers when bottom lines are fat. Companies generally have to go to the outside to hire, stealing people from other industries due to shortages of qualified personnel. These characteristics are usually present during the first third of the product/life cycle, signified by the first (left) vertical line intersecting the arc.

Conditions change as an industry matures. Promotions come more slowly, companies rich in profits become highly political and autocratic. People are hired from within the industry. Companies become selective in their recruiting. Arrogance abounds. (Whom the gods would destroy, they first make blind with power.)

The third phase of industrial evolution is represented by the area to the right of second vertical line intersecting the arc. This is the stage at which your job is most vulnerable—not because of you, but due to market conditions. Consolidations, reorganizations, plant closings, division closings, and selective cutbacks become necessary for survival.[2]

Most outplaced managers who make the transition from companies or industries that are late in the life cycle to younger companies have to learn new management techniques. Today's managers are "value added" managers. They are expected to contribute to the corporation in some real way and have a positive impact on profitability. In today's leaner, meaner, eye-on-the-bottom line service company, there is less room for managers who merely shuffle papers, attend meetings, and babysit a small group of workers. Today's manager is a working manager, a team leader who plays on the team.

THE BIG PICTURE

When we put all the pieces of the population puzzle together, we see an interesting, new picture of American economic life. We realize that the American economy is no longer one where most workers are employed by monolithic corporations engaged in the business of manufacturing goods, with an endless supply of young workers waiting to take the place of older workers, who happily retire in their late fifties and early sixties to fish and golf and garden.

Instead, ours is a service-producing economy, populated with vigorous smaller companies, where one worker with the help of new computer technologies completes tasks that it once took several workers to accomplish. It's a knowledge work force operating with fewer chiefs and more Indians, and supported by a burgeoning freelance, or contract labor, sector. In short, it's an ideal place for the older worker to thrive.

This fact is supported by research funded by the Carnegie Foundation and compiled in a book entitled *Our Aging Society*. Editors Alan Pifer and Lydia Bronte say this in their introduction to the book:

> Most men and women over 65 today are vigorous, healthy, mentally alert, and still young in outlook. In short, today's so-called elderly are not at all the same people as the elderly of

previous eras, even of two or three decades ago. And when the enterprising, aggressive, well-educated, baby-boom cohorts reach their "golden years," societal notions of the meaning of that time of life will change still further. Thus, old age in the aging society is a fluid concept invested with meaning, not by past expectations and norms, but by the characteristics of each generation as it reaches that stage of life.

    This tremendously important idea leads to . . . the notion that age 65 has become obsolete as a basis for life-course policy. A more realistic basis for policy might be a new concept we call the "third quarter of life," the years from about 50 to about 75 . . . Since the majority of Americans at the age of 50 today still have a third to a half of their life spans ahead of them, and since people between the ages of 50 and 75 will soon constitute nearly a third of the population, the third-quarter concept assumes that one's fifties and early sixties need not and should not be a period of gradual withdrawal and decline toward an inevitable cut-off point . . . Instead, these years should constitute a period of rebirth, with the awakening of new interests and enthusiasm for life, and new possibilities for being productive . . . More and more, an aging society will need the contributions its older citizens can make and, with long lives stretching ahead of them, they, in turn, will have an urgent need to stay productive.[3]

Ready to live the next third-to-half of your life? To get started on the next adventure? Good! Then let's go on to Chapter 2, where all the advantages the 50-plus worker brings to the work place are discussed. You will need to know because, in America in the 1990s, you still have to educate that person behind the hiring desk.

NOTES

    1. Sonnenfeld, Jeffrey, "Dealing with the Aging Work Force." *Harvard Business Review,* November-December 1978, pp. 81–92.

    2. *Professional Guide to Career Life Planning,* The Transition Team, 3155 W. Big Beaver Road, Suite 117A, Troy, Michigan 48084, pp. 10–11.

    3. Pifer, Alan, and Lydia Bronte, Eds., *Our Aging Society.* (New York: W.W. Norton, 1986) pp. 11–12.

2 | The Age Advantage

Bob E. became a client of The Transition Team at age 62. He had been plant manager of an aircraft engine parts manufacturer in Georgia for 20 years when his employer brought in a new manager, making Bob's decision to leave less than voluntary.

Bob had designed that plant, supervised its construction, and managed it from startup. As a result, his self-esteem was closely linked to that job, and all but lost with it. He and his wife were adamant about staying in the small southern town they had come to love, and where they had achieved status as "VIPs."

The Transition Team consultant who worked with Bob urged him not to foreclose his many options. He didn't see Bob as someone who would be satisfied with early retirement and the limited activities of small-town life. As a result, Bob reassessed his priorities. He realized how valuable were the depth and breadth of his experience. With a positive attitude about himself and his potential, Bob set about networking and in no time landed a position as vice-president of an aerospace company in New England.

Bob learned to use the age advantage to position himself as a vigorous, positive thinking, competent professional with a great deal to contribute.

At any age, looking for a new job can be a harrowing experience. No matter how skilled and experienced you are, the hiring process seems designed to destroy your self-esteem.

For the over-50 job seeker, it is even more so. The Age Discrimination in Employment Act notwithstanding, age bias still exists. It's unfounded, but it exists.

This chapter will arm you with the facts you need to overcome age discrimination—and maybe even your own attitudes about your age and what you think you can contribute—and get the job you want. By the time you read the last page, you will understand why your golden years are good as gold to any employer smart enough to hire you. And you'll be ready to persuade others, too.

Myths about the older worker abound. When used to justify discrimination against older workers, they hurt not only the older job seeker but potential employers as well. Let's look at some of the unsupported myths, and then contrast them with the well-supported facts that prove the value of the older worker.

MYTHS ABOUT OLDER WORKERS

1. Older workers are less productive than younger workers.

2. Older workers are absent more often due to illnesses associated with age. They are generally less healthy than younger workers.

3. Older workers don't contribute. They are just "putting in time" until retirement.

4. With aging comes a natural decline in physical strength, stamina, judgment, and ability to perform.

5. Older workers cost more to employ because their salaries and benefit costs (predominantly health insurance) are higher.

6. Older workers are rigid and inflexible. They don't learn new technologies nor readily accept new methods.

Fortunately, these perceptions aren't universal. A 1989 study by the well-known research firm Yankelovich, Skelly and White, Inc. for the American Association of Retired Persons found that, in general, older workers are perceived quite positively by American businesses—and this positive perception has increased in recent years. It is important to note, however, that in businesses with fewer than 1,000 employees older workers are perceived more positively than in larger companies.

A 1978 *Harvard Business Review* article detailed research findings that contradicted some long-held beliefs about the productivity and performance of older workers.

> It is important to explore how much factual evidence there is to support the stereotyping and the prejudices that link age with senility, incompetence, and lack of worth in the labor market. . . .
> Certainly, one does not have to look hard to find the elderly among the greatest contributors to current society. The list is long of older citizens who have made major contributions in all fields including the arts, industry, science, and government, and who continue to be worthy and inspiring members of our society.[1]

As for the myth that older workers are less healthy than younger workers, research has shown that while some physiological changes are indeed a result of aging, including loss of vision, failures in the immune system that lead to cardiovascular and kidney problems, and degenerative diseases such as rheumatoid arthritis—and 75 percent of those age 60 to 64 suffer from these age-related diseases to some degree—many of these problems can be controlled by modern medical treatment. Furthermore, studies show, *benefits* actually accrue from the natural process of aging.

> Reaction time seems to be affected by the increase in random brain activity, or "neural noise," which distracts the brain from responding to the proper neural signals. A fall in the signal-to-noise ratio would lead to a slower performance and increased likelihood of error. To correct for this possibility of error, performance is delayed to permit time to gain greater

certainty. Research . . . shows that older people are more scrupulous in the use of decision criteria before responding . . .[and] require a 75 percent chance of certainty before committing themselves, while younger people will take far greater risks.[2]

This is "science speak" for the simple fact that aging causes normal physiological changes, and we learn to compensate for them to keep our performance level up.

[R]esearchers . . . have found problem solving, number facility, and verbal comprehension to be unaffected by age. The ability to find and apply general rules to problem solving are more related to an individual's flexibility and education than to age.[3]

The health of older workers was also affirmed in the March/April 1990 issue of *Working Age,* a newsletter published by the American Association of Retired Persons. It reported the results of a study that showed it is the younger worker, not the older worker, who visits the doctor more frequently. Doctor visits decline for adults between the ages of 45 and 54 before rising to their highest levels for adults 65 and over, it reported. Adults between the ages of 55 and 64 registered 12.19 percent of all doctor visits in 1983, while young adults ages 15 to 24 registered 13.26 percent of all doctor visits.

In a 1977 survey, *Harvard Business Review* found that its readers perceived older workers as more rigid and resistant to change and thus recommended transferring them out rather than helping them overcome a problem. Researchers refuted this belief.

[C]hronological age never has been a valid means of measuring a worker's potential and now is illegal under the Age Discrimination in Employment Act. The strength of various faculties may slightly correlate with age in certain regards, but there is no categorical proof that age has an effect on capabilities.[4]

In fact, the article reported that older workers were found to perform better than younger workers in jobs requiring scholarship and artistic creativity as well as in sales, clerical, and manual jobs, to name a few.

> Age has had surprisingly little effect on *manual workers.* In several studies, performance seemed to remain fully steady through age 50, peaking slightly in the 30s. The decline in productivity in the 50s never seemed to drop more than 10 percent from peak performance. Attendance was not significantly affected, and the separation rate (quits, layoffs, discharges) was high for those under age 25 and very low for those over 45.
>
> These findings may not only indicate greater reliability among older workers, but also suggest that those who have remained on the job are, in some way, the most competent.
>
> . . . The need to evaluate potential on an individual basis, and not by age group, has been convincingly established in these studies.[5]

The same article described the success enjoyed by companies that never adopted mandatory retirement policies yet continue to operate profitably and efficiently with older workers. Among the companies cited were Ferle, Inc., a small company owned by General Foods, whose president at the time was 87 and whose workers averaged 71 years of age. "Older people are steadier, accustomed to the working discipline," claimed the chief executive.

Sales workers at Macy's department stores in New York have never had a mandatory retirement age and have demonstrated no apparent decline in performance attributable directly to age.

Bankers Life and Casualty Company, which retains top executives, clerks, and secretaries through their late 60s, 70s, and 80s, found that older workers show more wisdom, are more helpful and thorough, and perform their duties with fewer personality clashes. Studies on absenteeism at Banker's Life and Casualty show that those over 65 have impressive attendance records.

Polaroid found that older workers who chose to remain on the job after the customary retirement age of 65 tend to be better workers. Their theory is, "If you like to work, you're usually a good worker." Attendance among older workers at Polaroid is also "exemplary."

An article in the April 22, 1990 *New York Times* entitled "Too Much Retirement Time? A Move Is Afoot to Change It," discussed the benefits of hiring older workers.

> When Michael A. Leven, the president of Days Inn, began trying to hire retirees as reservation agents, he got no response to newspaper advertisements aimed at older citizens. The reason: Many had been rejected so often that they did not believe the ad.
>
> Mr. Leven was desperate, since his reservation agents at the time had a 30 percent absentee rate and a 180 percent turnover rate. "I knew they were out there, so we sat down and thought about where seniors gathered and sent people out to put up notices at senior centers," Mr. Leven said. "Then they began to believe that we wanted them."
>
> About 130 of Days Inn's 600 reservations agents are over 60, absenteeism is down to 3 percent, and the average time on the job is up to three years.[6]

There. I've shot holes in the first four of the six "Myths about Older Workers" listed earlier in the chapter. Overall, workers over 50 have low absentee rates, are in good health, and are found to be very productive and contributive to the work environment. They are praised highly for their work ethic and loyalty to the company. They are "players," not spectators waiting to retire.

As for the belief held by many companies that the older worker's salary is a higher expense, in fact the 50-plus group is no longer the highest paid. That distinction is held by 35-to-49-year-olds, reported the Yankelovich, Skelly & White study mentioned earlier. It showed that 35-to-49-year-olds have become more heavily represented in supervisory and management positions, which is in part due to the widespread implementation of early retirement programs. Health-care costs still remain a concern with most employers; but with all the benefits of hiring an older worker, employers definitely get "more bang for their buck."

The same study asked 400 employers—selected randomly to represent the universe of all companies with 50 or more employees—to rate older workers on nine selected characteristics. The following table summarizes some data on this point:

Characteristic	Percentage of Employers Rating Older Workers Excellent
Good attendance and punctuality	91
Commitment to quality	89
Solid/reliable performance record	87
Loyalty and dedication to company	86
Much practical, not just theoretical, knowledge	85
Someone to count on in a crisis	85
Ability to get along with coworker	79
Solid experience in job and/or industry	78
Emotional stability	71

In this study, however, older workers received low ratings on the following four characteristics:

Characteristic	Percentage of Employers Rating Older Workers Excellent
Good educational background	36
Physical agility	29
Desire to get ahead	27
Feeling comfortable with new technologies	22

In my own experience, I have found that these low ratings are almost always based on misconceptions. It is important to realize that such perceptions exist, however, so that you can be prepared to meet them head on and disprove them. Chapter 4 discusses exactly how to do that.

OPPORTUNITIES FOR OLDER WORKERS

Fortunately, research such as that we discussed in the previous section is beginning to have an effect on work place attitudes, and some progressive companies are tapping the potential offered by the older worker. Agencies that hire older workers almost exclusively for temporary assignments are beginning to spring up. They are placing workers in the limited clerical and accounting type functions typically associated with temporary employment. Many "executive temporary" employment agencies provide employers with senior people to fill managerial and professional slots on a short-term basis, at hourly rates up to the equivalent of a $100,000 per year salary.

The American Association of Retired Persons (AARP) (whose membership is open to all Americans 50 and over) maintains a computerized database on more than 100 older worker employment programs currently in place in private sector companies. Says AARP, "These . . . organizations have discovered that ability, not age, is the prime consideration when developing effective employment practices and motivated, productive employees."[7] Some of the companies and the programs each has designed are discussed next.

TEXAS REFINERY CORPORATION

Headquartered in Fort Worth, this company manufactures and markets building protectants and heavy-duty lubricants. It has an international sales force of 3,000, 500 of whom are in their 60s, 70s, and 80s. Hired as independent contractors, most work part-time to supplement social security and pension income. According to the company president, "TRC's older employees are more loyal and reliable and work harder than other workers."

ENERCOM, INC.

This is a company that, under contract to Georgia Power Company, administers a program called "Lending a Helping Hand." The program employs persons 65 years and older to weatherize the homes of their contemporaries. More than 275 older persons have weatherized approximately 36,000 homes since the program began in 1982. Georgia Power accomplishes its objective at reasonable cost and has strengthened its image as a progressive force in the community.

KELLY SERVICES

This nationwide temporary employment agency hires nearly 500,000 employees a year and has a special recruitment effort geared toward older persons and retirees because it found they prefer the flexibility of part-time and short-term assignments. Many employees in the company's technical division are retirees who return to work temporarily for the firms from which they retired.

THE TRAVELERS COMPANIES

Travelers' innovative "Older Americans Program" is a retiree job bank it started after a survey revealed many retirees wanted to return to work part time. The program has been so successful that the job bank is now open to non-Traveler's retirees in order to fill the demand by company supervisors.

To help employees sharpen their skills, the firm offers training courses on its newest office equipment. Retirees register with the job bank for fixed-schedule and on-call positions. Each day in the home office in Connecticut, about 250 of the 750-plus retirees in the job bank are at work. The company revised its pension system so that retirees could work as many as 960 hours per year without affecting their pension income and health-care benefits.

CONTROL DATA CORPORATION

This Minnesota-based manufacturer of computers also maintains its own job bank through a subsidiary, Control Data Temps. It employs a permanent part-time work force of 7,000, 20 percent of whom are professional-level employees. Employees can choose among a variety of programs. With "flextime," they can set their own schedule as long as it includes the mid-day hours of 10:00 A.M. to 2:00 P.M.. The "flexplace" option lets people work at home on computer terminals provided by CDC.

INTERTEK SERVICES CORPORATION

This California firm maintains a registry of 5,000 retired and semi-retired quality control engineers and technicians, some of whom are in their 70s. Recruited through professional journals and word-of-mouth, Intertek's workers are independent contractors who may accept or decline an assignment. They bill Intertek for their time and expenses twice a month, and Intertek sends the invoice to the customer.

F.W. DODGE COMPANY

Until recently, this Kansas-based firm hired temporary, part-time workers to transfer data from completed building permits onto forms supplied by the company. Because these temporary workers were not always dependable, the company redesigned the jobs as part-time, permanent positions, and recruited retirees to fill them. Ninety percent of available positions in 120 cities were filled by retirees in their 60s. The company found the older workers to be more reliable and the quality and accuracy of the data have increased.

TELEDYNE WISCONSIN MOTOR

A 1977 survey by Teledyne found that more than 50 percent of its skilled work force was eligible for retirement, and that spelled danger for the firm's productivity. Rather than start over with unskilled workers, Teledyne developed a landmark "Golden Bridge" policy to retain older workers.

Among the benefits offered, factory employees age 58 to 62 and salaried workers 60 years and older with 30 years service are automatically entitled to 160 extra hours of paid vacation. Those over age 62 receive 200 extra hours. The program also includes an additional $1,000 of life insurance for each year an employee participates, 60 percent of the basic pension benefits (a 5 percent increase) to surviving spouses of employees who work to age 65, and a one-third increase in pension benefits for each year a worker participates.

AEROSPACE CORPORATION

This Los Angeles company has 145 full-time regular employees who are 65 and older and also hires its own retirees as "casual" employees without disrupting their pension benefits. This program allows retirees to work up to 999 hours per year. Typically, they continue to work on assignments that had been theirs prior to retirement. Aerospace Corporation also developed a technical and administrative pool of retirees. The company values continuity on long-term projects and the "institutional memory" that results from its programs.

HASTINGS COLLEGE OF LAW, SAN FRANCISCO

Since World War II, when it experienced a shortage of seasoned educators, Hastings College of Law has recruited distinguished legal educators who have been forced to retire from other institutions. Older professors keep their pensions from all previous employment and, in addition to their salaries, are eligible for Hastings College of Law's insurance and annuity benefits. Eleven of the 51 full-time faculty at Hastings today are 65 or older.

FOCUS: HOPE

This small but successful organization in Detroit was created to train new machinists. Retired industry experts pass on their skills and knowledge to a new generation of machinists. In its first three years of operation, 45 companies hired graduates of Focus: HOPE. Pitney Bowes and AT&T also have programs that allow retiring workers to pass on their skills and knowledge.

McDONALD'S CORPORATION

The fast-food giant designed a special program, called "McMasters," for older workers. It provides skills training and job placement for persons who are 55 and older. McMasters participants are trained by a specially appointed Job Coach.

Once hired, the new employees begin a four-week training session under the guidance of a specially trained Job Coach. The pace and quality of the shoulder-to-shoulder training provided by the Job Coach helps McDonald's older workers to adjust to their new work environment.

ADVOCATING THE AGE ADVANTAGE

The American Association of Retired Persons (AARP) has been one of the most vocal and effective champions of the over-50 worker. According to AARP, companies that have shown a willingness "to ignore stereotypes and to look objectively at older workers . . . have discovered a vast source of talent and experience many of their competitors have overlooked."[8]

AARP offers a variety of educational and advocacy programs for older workers, who make up one-third of AARP's total membership. AARP's Worker Equity Department offers resources and technical assistance to employers interested in better using the skills and experience of older workers. Among resources available are:

∎ *Working Age,* a free newsletter highlighting employment trends;

∎ NOWIS, a database of employment programs;

∎ Management guides and training materials;

∎ Worker education materials.

Contact AARP, 1909 K Street NW, Washington, DC 20049 for information on materials and membership.

Along with AARP, many others know the older worker has a great deal to offer. It is heartening to see that many employers recognize that fact.

You don't have to target only those programs and companies that target the over-50 worker. You can compete head on with younger workers for any job. The rest of this book explains how.

But, if you're in the market for a new job involuntarily, even if your unemployment is due to forces beyond your control, it's going to take more than an armload of statistics to get you back on your feet and moving in a positive direction. For starters, it's going to take Chapter 3.

NOTES

1. Sonnenfeld, Jeffrey, "Dealing with the Aging Work Force," *Harvard Business Review,* November-December 1978, pp. 81–92.

2. Ibid.

3. Ibid.

4. Ibid.

5. Ibid.

6. Lewis, Tamar, "Too Much Retirement Time? A Move Is Afoot to Change It," *New York Times,* Sunday, April 22, 1990.

7. *Using the Experience of a Lifetime,* American Association of Retired Persons, 1988. (AARP, 1909 K Street NW, Washington, DC 20049)

8. Ibid., p. 39.

3 | Unemployed After 50: Coping Skills

Tom Z, age 54, was a production manager for a major manufacturer of machine tools. When his job was eliminated after 27 years, he was devastated. He experienced all the common reactions: denial, shock, fear, anger, hurt, humiliation, rejection, guilt, depression, and exhaustion.

Fortunately for Tom, he progressed through the stages of job loss reaction very quickly. And then he experienced relief—relief that a stressful situation had ended and that new opportunities awaited him. It didn't take The Transition Team long to convince Tom his new job was to find a job. He pulled himself up by the bootstraps, put on a happy face, and got right to work.

Tom called all the former management colleagues who had left his employer for others. One of his contacts put him in touch with a new U.S. venture being formed by a European company. Eight weeks after his termination, Tom signed on with the new venture as production manager.

Not every layoff or firing after 50 is identical to Tom's, but his story illustrates several commonalities.

THE "SACKED" CYCLE

This is my term for the emotional rollercoaster ride that often accompanies "getting sacked." Tom rode the rollercoaster once, got off, and got on with his life. Others stay on it far longer.

It is very normal to experience a full and disturbing array of human emotions in the wake of job loss. When those who are newly unemployed realize that what they are experiencing is normal and expected, it often helps them cope with their reactions, work through their feelings, and get on with the business of life.

Most people over 50 today were raised by parents who endured hardship and unemployment during the Great Depression. They were weaned on the notion that a good, secure job was a vital necessity. As a result, many older workers identify strongly with their work. Their sense of self, and of self-esteem, is rooted in their professional lives. Having launched their careers in the 1950s and 1960s when business was booming and adding layers of management, they came to regard termination as a dreaded stigma that occurred only as the result of incompetence.

That is not so today. As illustrated in the first two chapters, the world and the world of work are changing. Millions of people have experienced job loss in recent years. One survey of managers revealed that a previous "involuntary termination" is not held against job candidates today. Rather, employers are interested in the candidate's honest evaluation of why the termination occurred. More importantly, employers are impressed by candidates who are proactive. In other words, you stand a better chance of winning the race if you get right back on the horse.

I counsel many job seekers who are struggling with their emotional reactions and, consequently, can't orient themselves to get back into the race. Some candidates experience every phase of the cycle. Others skip over some of the phases. Eventually, however, most are able to work through their reactions and get back on their feet.

It is when a candidate gets stuck in one of the phases and can't work his or her way out of it that the job search is stalled and positive results delayed. Let's review the phases of the "sacked" cycle. Perhaps you will find yourself at one of these phases.

DENIAL

This happens even before you are terminated. You see people "like you" losing their jobs and saying "good-bye," but somehow you just know you'll be spared. You've always been a good performer (in your estimation). It can't happen to you. Then it does and you experience the next phase.

SHOCK

The stronger your early denial, the more severe your reaction when the ax falls. You may experience physical symptoms of nausea, chills, even fainting. You revert back to the denial phase. They have made a mistake. They didn't mean to fire *you*. It's a paperwork error. But there's your supervisor, looking a little wan, too, and assuring you that, unfortunately, your time has come.

Because you're not thinking clearly at this phase, it is vital that you don't take any action or sign anything until your judgment is restored. The executioner may very well be standing over you with a sheaf of papers for you to sign. But this is not the time to negotiate your outplacement benefits and reference letter contents. If you do, you may get less than you deserve.

Stay in control of the situation and your future security by telling your superior you need some time to think. Tell him or her you're going home and you'll be back the next day to collect your things and discuss the details of separation. You will be in a much better position to negotiate for severance pay, outplacement services, continuation of benefits, and a good reference after a good night's sleep. *Do not* act under pressure.

FEAR

Fired! Probably the strongest "f" word in the language. Even with a savings account and severance benefits, most people have financial worries—if not immediately, then down the road. Also, there is the fear of your family's and the community's reaction. "How will I ever face my friends?"

Loss of Control

Going to work every day and collecting a paycheck at the end of every week or month gives us a sense of *control* over our surroundings. In fact, we never do have real control over what happens to us, but we spend the major portion of our adult lives seeking control and believing we have it. Since it's very important to us to perceive we are in control, losing a job involuntarily becomes an even more unsettling experience than it should be. In fact, you have just as much control over your employment situation after you lose a job as before.

Anger

"Me? How could they do this to me? I've given the best years of my life to this company! I made them what they are today!" All these statements may very well be true, but if you focus on this phase of the cycle too long—if you stay angry—your anger and bitterness will be reflected in everything you do and say from here on out. You cannot conduct the best possible job search when you're obsessed with feelings of outrage and revenge.

My advice is to go ahead and feel that anger and rage. Don't take it out on your loved ones, and please, please don't demolish the office furniture or crack up your car. Okay, well, break a few dishes if it makes you feel better (but try to avoid the good china). Then transform that anger into positive power. With all that adrenaline flowing, you can be very persuasive if you have a grip on yourself, and you're likely to have the energy to accomplish a great deal of the early work on your search.

Yes, you did give them the best years of your life. But it was *your* choice and now yesterday is but a memory. It is *tomorrow* that should capture your imagination now.

HURT, REJECTION, HUMILIATION

A less noisy but nonetheless painful phase, this is anger turned inward and it's perhaps the most damaging kind. All humans thrive on approval and, work is where we get some of our strongest stroking. When that source of approval is temporarily removed, we may even blame ourselves. Blaming ourselves leads to the next phase.

GUILT

Like anger, you can get caught here. If you do, your demeanor will be that of a "whipped dog" instead of a barking, rabid one. You will feel worthless ("How could I have done this to my family? How will we survive? Who will ever want me again?) and you'll be completely unfit for telephoning and interviewing.

Some even feel they deserved to be fired, which is probably not true. This common reaction occurs among high achievers, many of whom are plagued with something psychologists call the "impostor phenomenon." They believe themselves unqualified for their positions and worry that someone will discover their inadequacies. When unemployment occurs, they naturally think their time has come and they've been "found out."

Go ahead, feel the guilt. For a day or two at the most. Then go away for a special weekend with your spouse and family or friends. Yes, treat yourself. *Believe* that you're worth it. You are. If you think you can't "afford" it right now, consider it an investment in your future. You need the peace of mind to be able to recover.

Besides, it will reassure your family that all is still well. They are bound to be worried, too. This is an excellent time to concentrate on your other roles (parent, spouse, child, civic volunteer), roles that perhaps have been ignored in the tense times leading up to unemployment. Try to have fun for a few days, refresh, and recharge. Take this book with you and do some planning for when you get back to the task of getting on with your career.

Work is only one part of yourself. Don't let the temporary loss of it destroy the rest of you. If you do, you'll find yourself in deepest depression, the next phase.

DEPRESSION

Like all the other phases, depression is a perfectly normal reaction—as long as it doesn't last. If you find yourself withdrawing, making fewer phone calls and scheduling fewer interviews, avoiding friends or spending a lot of time on the phone telling your story again and again to those friends, you must get help to get out. Getting stalled in this phase can stall your life. Remember the statistics from Chapter 2? You have a third to a half of your life left. Don't waste it.

EXHAUSTION

Physical, mental, and emotional exhaustion—complete, bone-weary exhaustion—is another common phase. It doesn't necessarily happen late in the cycle, after all the other reactions. It could happen immediately upon termination. If hard times in your industry or company have had you on the ropes for a while, this could be your *only* reaction to termination.

The best cure for exhaustion is a balanced diet, exercise, and a regular, active routine. Establish a regimen and stick to it. Set your alarm clock early, get up, exercise, shower, and dress in business clothes—whether you have appointments or not. (No kidding—you'll *sound* more professional talking on the telephone in a business suit rather than a jogging suit.)

Keep your regular appointments for haircuts, manicures, facials. If you don't have regular appointments, make them. Good grooming and outward appearance are just as important now as ever.

Don't stay up late watching movies because you don't have to go to work in the morning. You do have to go to work early—your job is finding a job or other economic pursuit.

RELIEF

If things have been really bad and you've been dreading the possibility of unemployment, when it actually occurs and you find yourself still standing and the sky still blue, you might actually have a sense of peace and purpose. "Okay, that's over. Now let's go on from here."

Even those who go through all the other phases of the cycle—from denial, shock, and anger through depression and exhaustion—will eventually come to this phase. It is a good place in the cycle. It means you've come to terms with the situation and are ready to get on with your life.

EXPAND YOUR MIND—EXPAND YOUR OPTIONS

Our friend Tom, mentioned earlier, went through the sacked cycle in less than a week. Then he got back in the saddle, and in no time he made several solid contacts and had a very good offer. Tom did something else very important, but something that many who are unemployed over 50 fail to do. He considered options he'd never previously considered.

Tom's salary at his "old" job was $75,000 a year. The new job offered a salary of $70,000. At first, Tom was reluctant to take a pay cut; but the employer was immovable on this ceiling. I encouraged Tom to negotiate for perks, which he did. He received the use of a company car, which was worth more than $5,000 per year.

The job also required Tom and his family to relocate from the Midwest to eastern Tennessee, which they were at first extremely reluctant to do. After investigating the region, they were soon enthusiastic about the lifestyle and lower cost of living it offered.

All too often after we've spent a decade or more in one industry, one town, one home, we get "settled." We are comfortable there, we don't want to move. We use our family's protests about their own sense of belonging to justify our own negative attitudes about change.

In fact, complacency and comfort are two of the biggest killers of both careers and lives. The making of an interesting life requires a willingness to explore and take risk. Before you reject a job or prospective job on the basis of salary, location, or title—investigate. The only impediments to success and happiness may be outdated ideas and an unexercised imagination.

This could be an ideal time to start a whole new career. Maybe you went to business school at the urging of your family when what you really wanted was to become a forest ranger. Or perhaps responsibilities to home and family from a young age prevented you from joining the Peace Corps or digging for lost civilizations.

Free from earlier financial and family constraints, many older workers are going back to school (where necessary) and preparing themselves for work they want to do and finally are able to do. Consider the experience of someone who lost his job and found a whole new meaning to his life:

> Lew L., age 60, was an expediter in a facility that manufactured mobile homes when the company moved and left him behind. Tired of many years of commuting and wanting to do something that involved people instead of products, Lew sought work in the human services field.
>
> Lew lives in a seashore community where there are many convalescent and continuing care facilities. He beat the bricks meeting all the managers of these facilities and expressed his sincere desire to spend the rest of his working years in human services.
>
> The director of a residence for the mentally ill took Lew at his word and hired him as an attendant. Although the pay was less than he'd been earning in factory work, the psychic and emotional benefits of the work (and the relief from the long commute) more than compensated Lew
>
> Lew's job performance was outstanding. Younger members of the staff came to depend on him for advice and counsel. He organized a resident council, which allows residents to participate in the management of the facility. After 10 years on the job, Lew "retired" *again* only to start work the very next day, in the very same place, but part-time.

Lew's story bears out the experience of many who learn that, when work is that rewarding, retirement pales by comparison.

If altruism isn't your cup of tea, perhaps fun and profit from self-employment are. Here are two stories of people who did just that:

> Neil G. was a machine builder for a manufacturer of machinery that produced roll steel. In the recession of 1980, when Neil was 55, he lost his job. He knew that the companies who used this machinery were laying off people, too, and that they would be without skilled service technicians to repair their machinery.
>
> Neil contacted every customer of his former employer and let them know he was available to repair their machines as an independent contractor. In no time, he was working 14 hours a day—as his own boss—and making more money than when employed. Eventually he received orders to build machines, which led to his getting his own shop. Now, at 65, he's busy and successful and employing others.

Neil responded to unemployment by assessing his skills and evaluating whether it was really necessary to find another job. He realized he could support himself doing work identical to what he had been doing, only now as his own boss. Many others can, too.

The next example illustrates a midlife career change that applied skills learned on the job to a whole new business:

> Joann N., 57, had sold advertising for a television station for 24 years when her job was eliminated. Not ready or able to retire, she bought a quick printing franchise. Her sales ability—combined with good franchisor support—made the business very successful. Joann's having more fun than ever and earning a good living while doing it.

If it's fun you're looking for, you might find it by making money at something you once did only on rare weekends when you could get away. For example:

> Jon and Joel W. are twins born in 1930. Both were very athletic. They lettered in varsity sports in college and pursued careers as physical education teachers and coaches. Eventually, Jon became a deputy superintendent of schools, Joel the admissions director of a state college.

Both retired in their mid-50s. Joel dabbled in a Christmas tree farm; Jon sold insurance. Although Joel liked being outdoors, watching trees grow proved boring. Jon made a lot of money selling insurance, but it was too much like work.

Today, both Jon and Joel are paid ski instructors in Utah and loving every minute of it. Both are very much in demand. Their maturity and teaching experience give them an edge over younger, less experienced instructors. They are staying fit and healthy and both can't believe they're getting paid to have so much fun.

All of these real-life stories illustrate the fact that you can have twice as much fun at 50, be twice as fulfilled, as you were at 25. In all of the situations described here, if unemployment had not occurred, most of these now happy and fulfilled people would have stayed in the same job, eventually burned out from boredom, and retired tired. Instead, they are vital and active, still learning and growing well into their sixties.

I guess then that the best "coping" skill I can give you is the realization that when one door closes, another opens—maybe even several others. In every cloud there is a silver lining and in every job lost, potential happiness found.

Once you've faced your feelings and regained your positive attitude, go on to Chapter 4. It is time to shape up, mentally, physically, emotionally, and occupationally.

4 | Take Ten Years Off Your Image

Dan C. had been a sales representative for an auto parts firm when he took early retirement at age 55. His modest pension benefits were supplemented and he made use of the outplacement services of The Transition Team.

In my first practice interview with Dan, I asked him to describe himself. He told me about his children and grandchildren. Then I asked him to describe his work. This was his reply:

"In the old days, you could take the buyer to lunch and clinch an order. Today, the buyer is likely to be a young gal with a technical degree. She's not interested in having lunch with an old man."

What's wrong with Dan's approach? Everything! Until Dan realized the outdated image he presented and learned to position himself differently, he didn't have much luck. Finally, he caught on, and today he's selling successfully again.

Age is a state of mind, one you can control. And you *must* control it when competing against younger workers for the best jobs. There are certain attitudes, actions, and appearances that are associated with "older" workers. (Read that: "too" old.) When you know what they are, you can avoid them.

Using Dan's case as an example, let's review the essential elements of updating your image.

TEN WAYS TO TAKE TEN YEARS OFF YOUR IMAGE

1. DON'T GIVE CLUES TO YOUR AGE.

"Grandchildren" is a dead giveaway. Of course, you're proud of your grandchildren. I am certainly proud of mine. You can decorate your new office with their photographs, but keep them out of your interview. Practice conversations that show what a vital, energetic, contributor you will be. Try to keep the conversation on work-related achievements.

If asked about your personal life, mention that marathon you're training for, the mountain-climbing vacation you just took, or your civic and volunteer activities. Avoid discussing the dates you graduated from high school or college, your Korean War service, the model year of the first car you owned, or anything that potentially could label you "over the hill." You know you're in prime time, and you have to convince the interviewer.

2. DON'T USE TIRED PHRASES.

Dan's language was peppered with worn-out, tired phrases that dated him: "In the old days," "old man." Here, from Jeff Allen's Best: *Win the Job* (New York: John Wiley, 1990, p. 101) are more "age old" expressions to avoid:

> "At my age . . ."
> "Back in the days when . . ."
> "Back then . . ."
> "In the good old days . . ."
> "It used to be that . . ."
> "Listen, son . . ."

"Nowadays . . ."

"Old timers like me . . ."

" . . . over the hill."

"The girls in the office."

". . . up in years."

"Way back when . . ."

"We used to . . ."

"When I was younger. . ."

"When I was your age . . ."

"When you get to be my age . . ."

"Years ago. . ."[1]

3. DON'T EXHIBIT PREJUDICE.

Dan was also exhibiting sexism in his attitude to women in the work force. In not so many words, he was saying, "What are they doing in a man's world?" Employers today tend to avoid hiring people whose behavior could prompt a harassment or discrimination complaint from a coworker or client.

Allen adds:

Don't call the interviewer "honey," or "dear." Don't refer to grown women as "gals" or "girls" and men as "guys" or "boys." Don't play into any prejudices or dislikes the interviewer may have.[2]

4. DON'T RELY ON OUTDATED BUSINESS PRACTICES.

Dan's attitudes and approach to his work were outdated as well. In an increasingly competitive work environment, you must be mentally and physically agile to succeed. Steak-and-martini lunches rarely are enough to earn their provider big commissions today. Success in almost every occupation requires awareness of the market forces affecting the particular industry and how they can be used to improve the company's bottom line.

Refer to Chapter 2 to the discussion of industry life cycles. If you started your career in a booming business that was on the upswing, and success came relatively rapidly, you may have to adapt your methods to a changing economy and the fluctuating fortunes of today's businesses.

5. DON'T FALL TECHNOLOGICALLY BEHIND.

Whether you're in engineering, accounting, marketing, communications, manufacturing, or another profession, knowledge and skills that keep your performance up to par shouldn't be ignored. Remember, in Chapter 2, the Yankelovitch, Skelly & White survey? Employers rated people over 50 poorly in technological competence and staying abreast of changes in their field.

The "old ways" aren't always bad, but they're not always the best, either. No matter how well you think they work, they'll do you more harm than good if the people who interview you, and those who evaluate you on the job, perceive that you're not keeping your skills and knowledge up to date.

Learning is a lifelong process. It is never too late to learn what you need to succeed.

6. DON'T CRINGE AT COMPUTERS.

This is the age of word-and data-processing. Even if your work does not involve daily computer use, computers affect most jobs.

Computers really do improve productivity, and taking the initiative to learn how to produce your own reports and memos, run spreadsheets, and develop proposals—or simply to learn how computer-generated information related to your job is produced—will prove you can change and adapt with the times.

Take advantage of any formal computer training your employer offers. If training is unavailable and you don't feel comfortable either asking your coworkers for help or trying to fumble your own way to computer literacy, get an inexpensive home computer and teach yourself at your own pace, free from the critical gaze of others.

If you can't afford to buy a computer outright, consider renting or leasing one. Personal computer rental is a growing industry. You may even be able to rent a model identical to the one on your desktop and in no time surprise your coworkers with your fluency. Many public libraries offer free use of personal computers, as well.

One of the biggest obstacles to computer use is the fact that so many workers—especially men—never learned touch typing. Check your local adult education catalogues for night courses in typing. The modest investment of time and money will be returned with interest.

America is straining under the burden of so much rapidly produced information, yet starved for *knowledge,* said John Naisbitt in *Megatrends* (New York: Warner Books, 1982). You have the seasoning and experience to take information and synthesize it into knowledge. But the first step in the process is access to the means by which the information is produced: computers.

7. DON'T MENTION YOUR AGE.

It is illegal for an employer to ask, and most won't. Instead, they'll try to pin you down through questions that give them clues. But you're onto them and you're not going to do that. Your age is unimportant if you're qualified for the work.

8. DON'T APOLOGIZE FOR YOUR AGE.

"You're probably looking for someone younger, but . . . "

"Who's going to hire someone my age?"

"I know I'm 55 but I'm a very young 55."

Candidates who open with lines like these get closed out of the job market. Apologizing for being 50 or older is self-deprecating, self-defeating, and plays into the prejudices that exist in the work place. Age shouldn't be a factor—and won't be—if you've followed the first seven rules—and the last two coming up.

9. STAY FIT AND HEALTHY.

This should go without saying, but it doesn't. While nearly everyone in America knows they should be eating right and exercising, and most people are at least somewhat aware of what constitutes a healthy lifestyle, many neglect this important youth-preserving rule. Ironically, it is often the over-50 worker—who can least afford to ignore it—who has paid attention to career, family, and other responsibilities at the expense of his or her health and wellness.

You need strength and stamina now more than ever to combat the stress of finding a new job. The temptation for many who find themselves out of work and less than perfect physically is to crash diet or begin a too-intense exercise program. Avoid such temptation; you'll just defeat your purpose.

Begin with a visit to your doctor for a complete physical. Ask about vitamin supplements, diet, and exercise. Start slowly and build gradually—but *stick with it.* Your efforts will pay off with more energy, vitality, and confidence.

In the meantime, you can compensate for a less-than-perfect physique with the right wardrobe, hairstyle, and grooming decisions.

10. KEEP YOUR WARDROBE CURRENT, BUT CONSERVATIVE.

A candidate's looks are one of the greatest sources of discrimination. You don't get a second chance to make a first impression. No matter how impressive your resume, how winning your telephone pitch—when you walk through that door the interviewer will size you up in seconds. If you don't make the grade, the rest of the interview is mere formality. It will be over in short order and you'll be on your way out.

There is a definite interview uniform, and those who wear it will pass that first muster. Without it, you'll never get a chance to state your qualifications.

There is no way to prove that you were discriminated against for wearing the wrong uniform. You probably won't even know. That is why it's best to dress conservatively and well.

A good guideline is to dress for the job you *want,* not the one you have. In other words, what are the people a few rungs up the ladder wearing? Size up their look—and copy it. In *Surviving Corporate Downsizing,* Jeff Allen explained why this works:

> Why do you think aspiring actors, politicians, lawyers, and people in every occupation look like the people at the top? Why do you think they're called "role models"? If you follow their lead, you'll succeed. Actors have known the rule for centuries:
>
> > Look the part,
> > and the part plays itself.[3]

Your interview wardrobe should be impeccable. You will need at least two suits or ensembles, because you'll be going on at least two interviews for every offer you receive. For men—as well as women—conservative, well-tailored suits with white shirts or blouses (commercially laundered and pressed), and ties/accessories to match are the best bet.

Solid navy blue and gray are the right colors. They are not only the accepted colors of the management uniform, but people who have studied the psychology of color say they convey power and authority. (Shades of brown and tan and other "earthy" colors do the opposite.)

For women, a dress with a jacket is also acceptable. Just make sure it is high quality, conservative, businesslike—and flattering. (Men—no sport coats or blazers, please.)

If you don't have two perfectly tailored, high quality suits or outfits that have been purchased within the past two or three years, go shopping. Quality doesn't necessarily mean high cost. Good brand name clothes are available at factory outlets or department store sales at reasonable prices.

If you're shopping at a department or specialty store, ask to talk with an "image consultant." Tell the consultant what you're looking for and consider his or her advice on styles, colors, and accessories best suited to you. Remember, shun fleeting fashion. Stick to the classic looks.

Shoes are also important. For men, perfectly polished black leather "wing tips" or other lace-up styles present the right image. If yours are worn, have them resoled and polished or replace them. Your belt and (if applicable) watchband should also be black leather. Women should wear dark leather pumps with a one-to-two-inch heel.

As for accessories, men should wear no jewelry except their wedding band and watch. (A school ring is also acceptable. Leave diamond pinkie rings and tie pins at home, however.)

Women's jewelry should be as simple and understated as men's. It should not be flashy or loud or it—rather than you—will get the interviewer's attention. The same goes for makeup. If it's obvious you're wearing it, it's overdone. Visit a skin salon or cosmetic counter in a better department store and tell them you want a "professional, businesslike look."

Women's handbags should be good quality leather in good condition, not too large, and of the same color as shoes. Handbags are not a necessity. Many women find they can store their keys, compact, lipstick, and comb in a compartment in their attaché cases.

(By the way, an attaché case is an interview necessity for both men and women. Keep in it a few extra copies of your resume, samples of any work you've done that will help at the interview, a pad of ruled paper for taking notes, and pens. If your old attaché is scuffed and shabby, treat yourself to a new one.)

When you go shopping to update your work wardrobe, keep in mind your total look. If you're shopping just for accessories, wear the clothes, or take them with you, so that you can try on the entire outfit. If you're shopping just for the suit, don't be afraid to take your own shirts or blouses, ties, and other accessories. This is a big investment. It has to be right.

Once you have it all put together, spend some time in front of the mirror. Look at yourself critically. If you cannot be objective, take someone along who can. (Someone who understands and appreciates the interview uniform.) Does it fit well and flatter you? Are there any jarring notes? No shiny spots? No frayed cuffs? Does the image in the mirror convey confidence, vitality, and authority?

The only time you can bend these wardrobe rules is if the job you are seeking is in a high fashion or entertainment profession where people tend to show creative expression through their clothing. I said bend—not break. In such cases, a more colorful tie or scarf would be acceptable. Don't abandon the interview look. People expect it.

Also remember that human resources people, even in creative and fashion professions, tend to reflect the corporate look. If you're unsure, err on the side of caution and keep it conservative.

As for hairstyle, I endorse Jeff Allen's advice, also from *Surviving Corporate Downsizing,* to keep it simple, attractive, and conservative—like the uniform:

> Is [your hair] limp, listless, and lumpy? If so, there are (at least) a jillion shampoos and conditioners on the market that are intended for *daily* use.
>
> What about the style? Men let their sideburns grow and don't shave the back of their necks between haircuts. They use 50's goose grease because they've always used it. They think hair spray will stimulate the production of female hormones. They tape, weave, and patch their hair where nature has come and gone.
>
> Women let their hairdressers try out their latest creations on their heads. Neon birdnests are the result. Half the dyes, tints, and rinses are designed to reduce the effects of the other half. This should tell you something about that nonsense.[4]

You don't have to "wash away the gray." It depends on how it makes you look. Are you more distinguished as a result? Or are gray streaks here and there less than flattering?

Toupees, wigs, and dyes that look less than natural are a no-no. So is men's hair that touches the collar and women's that touches the shoulder. Women with long hair should wear it up in a smooth, natural style.

These ten tips are not meant to regulate your personality or conform you into a coma. They are created out of a recognition that you're competing in a tough market, and you want the best job you can get. You're competing against others—both younger and older—who are observing these rules. You're being interviewed—and accepted or rejected—by decision makers who understand the same rules and adhere to them.

While society is full of individual acts of brilliance, creativity, and style, in general the occupants of the business world are creatures of habit. The habits keep changing, but anything that differs from the current norm is frowned upon.

Make sure your habits, attitudes, and outward appearance are in step with what is accepted out there, and you'll be in where you want to be in no time.

Now that the preliminaries are out of the way, it's time to begin Part II, The Preparation. Chapters 5, 6, and 7, will sharpen your skills in planning, targeting, and researching for the right job. Chapter 8 will help you prepare to defeat age discrimination.

NOTES

1. Allen, Jeffrey G., *Jeff Allen's Best: Win the Job,* New York: John Wiley, 1990, p. 101.

2. Ibid., pp. 101–102.

3. Allen, Jeffrey G., *Surviving Corporate Downsizing,* New York: John Wiley, 1988, pp. 69–70.

4. Ibid., p. 70

PART
II

The Preparation

5 | Taking Stock of a Lifetime

What does the 50-plus worker have that younger workers don't have? Experience! Lots more experience, at work and at life. Here's what one employment industry insider says about the value of that experience:

> The last 50 people we've seen who are over 40 have gotten better jobs despite the fact they were *obsessed* (it's fair to say possessed) by fears that they'd never work at a decent job again. Logic-defying as it seems, *work experience is far more important than potential right now.* Companies are in the most risk-averse period in recent memory, going all the way back to the Great Depression, and they feel more comfortable paying higher salaries to people who come with some *guarantees,* i.e., experience.[1]

This is good news for the over-50 job seeker, and welcome relief from the conventional wisdom that says "you're only as good as your last accomplishment." With the work place emphasizing only what's new and novel, older job seekers are usually warned to cut their resumes from the bottom and reveal only the most recent 10 or 15 years' achievements.

As discussed in Chapter 9, your resume should be limited to the past 10 or 15 years of accomplishments. That's because employers see your resume before they see you, and you don't want to give them a chance to discriminate against you on the basis of age. But that doesn't mean I want you to *forget* everything that happened more than 15 years ago. *All* your experience and achievements—both in and out of work—are precious gems to be mined, polished, and displayed to prospective employers.

Your achievements build on each other, and build *you* into the qualified, competent, and confident career professional you are today. Therefore, everything you've done has validity. Even your failures are important. They taught you what *not* to do. And, while you may not list every experience and accomplishment on your resume (where would you get enough paper and ink?!), your first step in a successful job search after 50 is to take all that experience out and look at it carefully. Inventorying your skills and accomplishments will not only give you confidence (Hey, I've really done a lot with my life!), but knowing where you've been will also help you decide where you want to go.

OBJECTIVES

As you set out to inventory your skills, keep in mind the objectives of this task. A skills inventory serves to:

1. Provide an organized framework for thinking about your skills and abilities, which helps prepare you for successful interviewing.

2. Catalog all skills and interests as a base upon which to develop the widest possible range of career/life options.

3. Provide the raw material to develop a succinct, yet powerful, resume.

4. Eliminate lost time and wasted effort caused by a random search.

THREE PHASES

The inventory process consists of the following three phases:

1. Developing a complete work history of both paid and unpaid posts.

2. Itemizing and describing accomplishments in all aspects of your life, including work, family, hobbies, civic and volunteer work, and so forth.

3. Compiling a list of skills according to category (e.g. technical, interpersonal, administrative, managerial, leadership, etc.).

PHASE 1: WORK HISTORY

Starting from the present and working backward, list each and every job you've held, including the dates of employment, title, and company name and address. Include any positions you held outside of work (local government, trade associations, committees, etc.) or volunteer organizations you served. Be sure to list positions within organizations such as committee chairperson, fundraiser, event organizer, and so on.

To simplify your inventory, copy and use the "Work History" form I've provided, (Figure 5–1) or a similar one. List *all* jobs held, including part-time work during high school and college, internships served, and military experience. If you have held more than one job for the same employer, list each job separately. You will likely need more than one page to complete your work history.

Even summer jobs and part-time work during high school? you might ask. How could a summer job as a store clerk be pertinent to the job search of a CEO? Everything you have done tells something about you. Were you working to save money for college? That's evidence of setting a goal and achieving it. Did your earnings help you buy a car or pay family expenses? That shows that from an early age you had a strong sense of responsibility.

Don't leave anything out—and don't rush through. This job search phase is too important to do haphazardly or incompletely. It creates the foundation that the rest of your career will be built upon.

Block out a half—or even a whole—day on your calendar, find a quiet place free from interruption, and relive your

Figure 5–1 Work History Form

WORK HISTORY

Dates of Employment		Job Title	Company & Address
From	To		

1. _____ _____ _____ _____

                                                                            _____

2. _____ _____ _____ _____

    _____

3. _____ _____ _____ _____

    _____

4. _____ _____ _____ _____

    _____

5. _____ _____ _____ _____

    _____

6. _____ _____ _____ _____

    _____

7. _____ _____ _____ _____

    _____

8. _____ _____ _____ _____

    _____

9. _____ _____ _____ _____

    _____

10. _____ _____ _____ _____

    _____

occupational past. Go through old calendars, outdated re-sumes, files, newspaper clippings, company newsletters, award certificates, annual reports, address books—anything that will aid you in remembering where you've been and what you achieved there.

The work history phase produces only a list of all jobs held. Next you must examine closely each job in your work history for *accomplishments.*

PHASE 2: ACCOMPLISHMENTS

Your past accomplishments—not your assigned responsibilities—are the key to your successful search. Prospective employers want to know less about job title and functions than they want to know about your achievements within the job. Don't assume a prospective employer will automatically understand what an "engineering manager" or a "customer service representative" does. Positions vary widely in duties and relative importance within organizations. You are preparing to communicate in your resume and interview not what *jobs* you've *held,* but what *skills* you can *offer*—and you must demonstrate your competence with facts and figures.

As long as you're searching your memory bank, this is also a good time to identify potential references and sources of job leads. Use a form like that shown in Figure 5–2 for *each* position—paid or unpaid—listed on your Work History.

Under "Potential References/Job Leads," list the names of people you worked with on that job. It could be a supervisor, coworker, vendor, or customer. If you know their current telephone numbers, include them as well.

Under "Brief Description of Responsibilities and Span of Control" summarize your job functions, the number of people reporting to you (if any), and their job titles or categories.

Under "Significant Accomplishments" describe how your ideas and actions solved problems, saved dollars, improved productivity, and so forth. Use the following list of some customary areas of accomplishment, taken from The Transition Team's "Professional Guide to Career Life Planning" as a guide.

Figure 5-2   Position Analysis

POSITION ANALYSIS

Job No.: _____

Company Name: _____

Address: _____

Telephone: _____

Title: _____

Dept./Div.: _____

Dates Job Held: From: _____   To: _____

Reported to: _____   _____
                          Name   Title
Potential References/Job Leads & Current Business Tel #:

_____

_____

Brief Description of Responsibilities and Span of Control:

_____

_____

Significant Accomplishments (Use dollars and numbers where possible)

1. _____

2. _____

3. _____

4. _____

5. _____

6. _____

Special knowledge and skills used (or learned as a result), such as foreign language, computer programs, accounting systems, etc.

_____

_____

_____

Areas of Accomplishment

Trained others (What results? Why were you chosen?)

Received awards (For what? From whom?)

Instituted new procedures (Again, with what results?)

Installed new machinery/systems

Increased sales (How? How much? Use $$ and %.)

Saved the company money (How? How much?)

Performed work outside job description (What? Why?)

Identified problems others did not see (What were they? How were they solved?)

Improved productivity or efficiency

Increased employee morale

Enhanced customer service

Reduced waste

Improved quality of company's products or services

Identified and tapped new markets (What were they? How did you tap them? What were $$ gains as a result?)

Reduced inventory levels

Achieved purchasing economies

Invented new products

Organized a fundraising effort (How much raised? What percent of goal?)

Exceeded sales objectives (By what percent?)

ACCOMPLISHMENTS OUTSIDE OF WORK

In addition to accomplishments in your paid and volunteer positions, describe achievements in your hobbies or leisure pursuits. Did you learn to ski (hang glide/surf) at a late age? Have you run a marathon? Did you take part in an expedition or safari to an unusual place? Have you designed and built anything (from a tool shed to a vacation house)?

Your goal here is to mine the riches of your past to produce the best possible career presentation. List *everything* that highlights your accomplishments as well as your proven ability to learn new skills and remain a vital, active participant in life both in and out of work.

PHASE 3: SKILLS BY CATEGORY

Everyone has skills in more than one category. Even if your entire career has been in manufacturing management, you've undoubtedly developed strong "people" skills as you hired, trained, supervised, negotiated, and motivated the people in the plants.

From the accomplishments listed on your position analysis forms, create a third—and final—list. This one is a list of skills according to the category of job they fall under (for example: engineering, sales, technical, information systems, accounting, personnel) along with the appropriate accomplishments (using numbers, dollars, and other concrete figures) wherever possible to document those skills.

Use a form like that shown in Figure 5–3. At the top of the page, list the category. In the left-hand column, list the skills that fall into that category. To the right of the skill, put the accomplishment (or accomplishments) that document that skill.

As you progress through all three skills inventory phases, note those accomplishments that make you especially proud, and of jobs (or parts of jobs) that you enjoyed more than others. It might have been the kind of working environment, the company "culture," or simply the types of duties that pleased you.

At the same time, note what you did *not* like. What kinds of work bored, frustrated, or angered you? What memories rankle when unearthed? As an over-50 job seeker, you need to know your past likes and dislikes for a very important reason. Mistakes and career detours are for the young. This time around, you're going to target a job you really want. Chapter 6 discusses how.

NOTE

1. *Kennedy's Career Strategist,* Vol. V., No. 2, February 1990.

Figure 5–3 Skills by Category

Skills by Category

Category: _____

Skill	Documentation (Accomplishments)

6 | Targeting the Right Job

There is only one success—to be able to spend your life in your own way.

Christopher Morley

Perhaps when you were younger, just starting out in your career, you didn't feel you had the luxury of *choice* in a job. You took what you found available, and your career took off from there. Perhaps you were happy about it. More likely you became adjusted over time. It wasn't what you had dreamed about, but it wasn't so bad, either. It was a good living. You bought a house, maybe a boat or a vacation retreat, put the kids through college. No, not a bad life at all.

But, admittedly, it's been something less than complete fulfillment—that deep down satisfaction that comes from doing work you love and believing that what you are doing is truly *worthwhile*. The kind of work that makes you spring out of bed in the morning eager to start the day, and doesn't take a few mixed drinks and the newspaper to "wind down from" at the end of each day.

Wouldn't it be nice to have a second chance at your dreams? It is not too late.

Most—although not all—50-plus job seekers are in a better position to take time to find the *right* job than they were 30 years earlier. For many, family responsibilities have eased—

and with them the relentless drain on the family purse. For couples, there might even be two incomes and a paid-up mortgage. Even when there are college tuition payments to meet, most kids are perfectly capable of working at part-time jobs to help with their own expenses. You would be amazed what *your* kids are capable of doing.

HOW OLDER WORKERS ARE REDEFINING WORK

With a comfortable cushion of cash, either from their own savings, company-sponsored savings plans, a spouse's income, or severance pay, many over-50 job seekers have the means—and the desire—to redefine their goals in life and locate work that, for the first time, is something they *want* to do.

- They are working part-time, and spending more time on leisure pursuits and hobbies.
- They are self-employed consultants and subcontractors providing services to companies or even their former employers, but now they're naming their fees and hours.
- They are working full or part time at unpaid, volunteer positions to help create a better world.
- They are working as part of the new wave of executive and professional "temporaries," which allows them to structure work assignments around other interests.
- They are starting businesses in fields wholly unrelated to their "first" careers.
- They are turning former hobbies into lucrative sources of income.
- They are relocating to different parts of the country— even the world—and starting all over.
- They are returning to college and learning new skills and professions.

In short, they're taking a *risk* and reaping huge *rewards*. Most of them will tell you with exuberance that they are happier than they've ever been. This really is *their* time. For these fortunate 50s, the golden years are just that.

Is there something you've always wanted to do? Start your own business? Turn a woodworking or fashion design hobby into a paid profession? Even something as simple as moving to a better climate—or closer to the sea or the mountains where you can pursue recreational activities you love?

What's stopping you? You *can* take charge of your life—and change it for the better now. First, however, you must set a goal and develop a plan for achieving it. This chapter is designed to help you select the goal you want to achieve.

Although I'm in favor of risk, however, I'm opposed to unreasonable risk. There are ways to achieve the upside of challenge and fulfillment while minimizing the downside risks of financial exposure and lifestyle upheaval. You don't want to sell your farm in the Midwest and buy a boat to live on, only to find out afterward that you get hopelessly seasick and you're allergic to seafood. Here's a true story along similar lines:

> Larry and Joan H. were both in their 50s when, with kids grown and gone, they decided to retire from their separate careers, buy a 50-foot sailing vessel, and set out to see the world. Fortunately, the wisdom and judgment they'd gained from their successful management careers kept them from selling their house in New England or using up all their resources on their boat purchase. And, also fortunately, they bought sufficient insurance coverage.
>
> One stormy night early in their voyage, their anchor line broke and they found themselves dashed again and again against the rocky Maine coast. The night was a struggle forever etched on their memories. When morning came, they believed themselves lucky to be alive.
>
> Although the boat was beyond repair, their dreams were not. They returned home and, from the horrors of that experience, re-evaluated their dreams. They decided they still wanted to spend the rest of their lives in some common pursuit, but their narrow escape had made it equally important to be near their children, all of whom had moved to the West Coast. So they devoted another year to research and a search for the right situation.
>
> Today, they own a beautiful home on Puget Sound and a bookstore in a small town near Seattle. (They really enjoy the 30-minute ferry ride to work in the morning.) They were able

to buy more for their real estate and business dollar in the Northwest than they could have in their native New England. They see their children and grandchildren regularly, and the pace of their work lives is far more to their liking than when they toiled for big companies. (But not quite as breathtaking as a night on an angry sea—which is just fine with them.)

MINIMIZING RISK, MAXIMIZING REWARD

Larry and Joan's story has a happy ending. It might not have been so. Perhaps they needed to experience almost losing their lives at sea to discover they really didn't want to spend their lives on the sea after all.

The decisions you make now could mean the difference between happiness and mere "adjustment," living fully and just existing. To ensure you achieve the former, seek your rewards but cover your risks. Don't mortgage everything to invest in a business—yours or anyone else's—until you've thoroughly investigated the operation. Don't join the Peace Corps and move to Africa until you've *been* to Africa and are reasonably certain you can tolerate the change. And always operate from a reasonable cushion of resources. Have something to "come home to" as Joan and Larry did, if at all possible.

Whether it's a complete life change you seek, or merely a more satisfying version of your existing career, this phase in your search requires you to decide what you want to do with the rest of your life. Now is the time for targeting. After describing the step-by-step process of targeting, I'll give you more information about options such as self-employment, part-time, temporary, and volunteer work.

KNOW YOUR PAST, FIND YOUR FUTURE

The past *is* your key to your future. You completed the first step in targeting the right job (in the worksheets in Chapter 5) when you highlighted in your achievements and skills inventory what you enjoyed doing most.

What did you like about certain jobs? What didn't you like? What achievements gave you the most satisfaction? Complete the checklist in Figure 6–1 to construct a picture of the perfect job for you. Put a check to the left of all true statements. Write yes in the blank to the right of each statement you checked if you *honestly* believe you're also good at it.

Now, review your checklist. Summarize it. Did you check off any items that you *like* to do that you then answered "no" to the question "Am I good at it?" This might indicate that you simply need more training to be good at something you'd like to do. If so, explore educational and training opportunities. The next section will help you appraise honestly what strengths—and weaknesses—you bring to your search.

Figure 6–1 Target Job Checklist

TARGET JOB CHECKLIST

Do I like to:

Am I good at it?
(yes or no)

☐ Work with people as part of a team? _____

☐ Work alone? _____

☐ Work with numbers? _____

☐ Work with ideas? _____

☐ Be in charge? _____

☐ Not have the responsibility of management? _____

☐ Create something tangible? _____

☐ Work within a more conceptual framework? _____

☐ Administer and manage? _____

☐ Have a variety of things to do at all times? _____

☐ Do one thing at a time until it's finished? _____

Do I like to:

Am I good at it?
(yes or no)

☐ Travel on business? _____

☐ Avoid business travel? _____

☐ Have a regular salary and benefits? _____

☐ Earn more through commissions and
other compensation incentives? _____

☐ Supervise and motivate others? _____

☐ Work with customers and/or the
public? _____

☐ Sell? _____

☐ Solve problems? _____

☐ Work 40-plus hours? _____

☐ Work fewer than 40 hours per week? _____

☐ Work only traditional hours? _____

☐ Work flexible hours? _____

☐ Work in a large company? _____

☐ Work in a small organization? _____

☐ Work with an established company? _____

☐ Work for a young, on-the-move
employer? _____

☐ Learn new methods and skills? _____

☐ Stick with what I know best? _____

☐ Write? _____

☐ Speak in public? _____

☐ Work with computers? _____

☐ Do physically active work? _____

☐ Be my own boss? _____

☐ Let someone else have the headaches
of ownership? _____

PERSONAL CHARACTERISTICS

In completing the Target Job Checklist, you ascertained what you like to do, but you also evaluated honestly if you were skilled at it. What are your strengths and weaknesses? Are you good at solving problems with processes, but not good at making others understand what has to be done? Are you a leader or a doer—or both? Do you make dynamic presentations or pale at the thought of speaking in front of others?

Referring to your skills inventory from Chapter 5 and the target job checklist you just completed, list your strengths and weaknesses in Figure 6–2.

Figure 6–2 Personal Characteristics

PERSONAL CHARACTERISTICS

Strengths	Weaknesses
_____	_____
_____	_____
_____	_____
_____	_____
_____	_____
_____	_____
_____	_____
_____	_____
_____	_____
_____	_____
_____	_____
_____	_____
_____	_____
_____	_____

EVALUATE THE ENVIRONMENT

The next step is to target the kind of surroundings you seek. What, if anything, would you change about the job environment you're accustomed to? Would you prefer to live in a different climate? Would you like to move from an urban to a rural setting, or the reverse? Would you like to turn in your sales bag for a desk? Vice versa?

If you've worked in big, corporate bureaucracies most of your life, you may long for work in a smaller organization where you have access to top management and the ability to respond quickly to market changes before half a dozen committees have approved a strategy. Or, you may like your work and industry just fine and all you want is a change of scenery. Consider the following true story:

> Jay M., was an accountant in a cold northern city. His "desk job" occupied him some 60 hours per week for more than 30 years. He had always had a weight problem, which worsened with age, and he wished his work were less sedentary.
>
> When Jay took an early retirement package at 55, he moved to the Southwest to become the accounting manager for a group of three resorts. He still works up to 60 hours per week at a desk, but he also puts in about 20 hours per week, free of charge, in the resort's gym, on the golf course, and in one of several swimming pools. He has lost 35 pounds and is healthier than ever.

Jay's job didn't become "less sedentary." The circumstances surrounding it did. Rather than change his career, he merely changed its environment.

Smaller company? Bigger company? Relocate to a different part of the country? A different kind of work? These are some of the questions you will answer when you complete the "job environment summary" in Figure 6–2.

List all factors of job environment you seek, even if they seem unimportant to write down. Include everything you want to find in the internal environment ("position" details, including type of work, physical surroundings, people you'll work with, salary and benefits) as well as the external environment (company, industry, and geographical location).

Be realistic. The more closely your goals are aligned with the real and irreversible trends taking place in the world of work, the more likely you are to get what you want. Consider a range of factors, including:

THE POSITION

The United States Department of Labor publishes the *Occupational Outlook Handbook,* which lists occupations with the greatest predicted growth. The June 1990 issue of *Money* magazine, in an article entitled "Work in the 1990s" profiled 15 of these, including:

Bankruptcy lawyer. The people who clean up after financial failures profit tidily from misfortune. Average salary: $75,000 to $150,000. Partners earn $200,000 and up.

Operations research analyst is the 90s name for what was once called "efficiency expert." This position involves sophisticated computer analysis of an organization's daily activities to determine the most effective way to allocate human and financial resources. With a bachelor's degree in math, computer science or engineering and a master's degree in operations research or management science, these new professionals earn on average $50,000 a year. Top performers can earn $150,000 and up.

Human resource manager. With the dwindling numbers of entry level job applicants, companies will be challenged to find enough workers to fill certain available positions. The old "personnel officer" is now accorded a new title and, with it, new respect for a tough job. Salaries are going up, too. Middle managers in the 90s will earn from $40,000 to $70,000. Top executives: up to $165,000. A liberal arts degree along with an M.B.A. are the best educational qualifications to offer.

Other growing professions cited by the *Money* article were health care cost manager, management consultant, environmental engineer, physical therapist, geriatric specialist, teacher, and registered nurse. Check the *Occupational Outlook Handbook* in your local library to determine where your current—and, if applicable—future career choices rank.

THE COMPANY

Bigger, older, established companies are not the best targets for over-50 job seekers. Few *Fortune 500* companies are actively recruiting in large numbers. Often, America's "blue chip" employers are downsizing through attrition and other, more aggressive staff-cutting measures.

Most of the jobs available today (up to 500,000 openings *every month*) can be found in smaller, newer companies. These are ideal openings for older workers whose experience in the older organizations is valued. And, in a smaller, less bureaucratic organization, you can gain access more readily to the hiring decision maker. You'll have a greater opportunity to use your persuasion skills to overcome any disqualification based on age.

In addition to the type and size of company, its management philosophies, corporate culture, your own job description, salary and perks, you must target the industry.

Remember earlier cautions to avoid "buggy whip" industries. A more current example can be found in the publishing industry. While traditional typesetting companies have hit hard times, "desktop publishers" are doing a booming business working on simple, relatively inexpensive computer equipment that didn't even exist in 1980. They are able to produce printed materials in less time and at reduced cost than the older method known as "photo composition."

What emerging trends do you see fueling the economy of the next decade—and century? For greatest opportunity and satisfaction, transfer your skills and experience from the old to the new. The axiom that "you can't teach an old dog new tricks" simply doesn't apply to the over-50 worker. If I believed it did, I wouldn't have bothered to write this book.

GEOGRAPHICAL LOCATION

Many jobseekers I've counseled have told me, "I'll only relocate if I have to, if there is no other choice." Others have been so eager to relocate that they neglected, despite our urgings, to fully evaluate the move and make sure it was a good one.

In the late 1970s and early 1980s, many people laid off from automotive industry jobs in the Midwest heard about opportunities in Texas, and the resulting migration was second only in size and eventual disappointment to the 1930s migration from the dustbowl to California. Because climate, lifestyle, and other environmental factors were too dissimilar for many—and because Texas, in its turn, experienced its own economic crisis—a "promised land" became a nightmare. Although there are no exact statistics, I'd venture to say the majority of those transplants have found their way back to their more familiar Midwestern roots and to satisfactory, if different, jobs.

That was an example of leaping before looking, something that most readers of this book cannot afford to do. To thoroughly evaluate whether to move to a different locale, first identify those characteristics that you enjoy and appreciate in the area where you now live. Be objective about desirable characteristics your home now lacks. Include in your evaluation:

■ *Economic Factors*
   Cost of housing
   Food prices
   Average household income
   Overall cost of living
   Taxes (state and local)
   Personal income growth
   Job expansion

■ *Housing*
   Percentage of listings sold in past six months
   Number of listings on the market
   Market appraisal value of your home
   Potential rental income from your house if not sold
   Best months to sell
   Average length of time homes remain on market before sale
   Utility costs
   Property taxes

■ *Climate and Geography*
   Advantages (what you like)
   Disadvantages (what's missing that you'd like to have)

▪ *Education*
Educational resources for children, if applicable
Educational resources for adults

▪ *Health Care and Environment*
Air quality
Water quality
Proximity to toxic wastes and other unhealthful conditions
Access to health care
Crime

▪ *Arts and Leisure*
Recreational facilities
Professional sports facilities
Art museums
Libraries
Theater
Symphony
Opera company
Dance company
Other

▪ *Transportation*
Public ground transportation—buses, subways, and so on
Airports—distance from major airport
Commuting time/distance to work

There may be factors about your current home that you take for granted. You might move to another city only to discover that your teenagers can't get to jobs and other activities unless you buy them a car. Suddenly you miss that great city bus system in your old home.

You may opt for a more rural location and find that your frequent business travel is complicated by a one-hour drive to the airport and inconvenient connections to larger airports.

You may think you're getting a dream home with acreage, and then discover a toxic waste dump behind the trees in your backyard.

No such pitfalls will trip you up if you do your homework. First decide what you want you want to find in both the internal and external environments, then write a paragraph describing your target position, company, industry, and geographical location on the job environment summary form (Figure 6–3). Chapter 7 will list resources for researching the perfect job, company, city, state, and even country for living the next phase of your life.

Figure 6–3   Job Environment Summary

JOB ENVIRONMENT SUMMARY

Internal Environment

The position

_____

_____

_____

External Environment

The company

_____

_____

_____

The industry

_____

_____

_____

Geographical location

_____

_____

_____

Now look at all three worksheets from this chapter. Based on what you listed in them, write a paragraph describing your target job, company, and locale. Your target job(s) should be a pleasing mix of what you do best and what you like best. (Usually, these are one and the same.) Describe the kind of work, the kind of company, industry, geographic location, hours, pay scale, benefits—even the number of windows in your office and the menu in the company cafeteria. The more specific, the better.

Because the more clearly you see your target, the more likely you are to hit it. Chapter 7 will help you locate the companies—and decision makers—with openings that fit your objectives.

In the event you are among the 20 percent of older workers who don't really want another job, but would prefer to take a different tack with the rest of their lives, for the remainder of this chapter I discuss the pros and cons of some alternatives to getting another job. Let's look now at self-employment, part-time employment, temporary assignments, and volunteer work that leads to paid positions.

SELF-EMPLOYMENT

In the business shakeout still being experienced in America, many workers are leaving corporate payrolls to become consultants or "independent contractors" providing services to their former employers. For some, it's a gold mine of unlimited income and career satisfaction. For others, it has all the appeal of being an itinerant laborer in a business suit. Consider the benefits as well as the disadvantages before you print your business cards.

BENEFITS

The two main benefits of self-employment are attractive, indeed. They are (1) self-determination and (2) the chance to greatly increase compensation.

Independent contractors are free to offer their services to many different companies. They often earn more money than they did as employees. They make their own hours for work and play. Although many find they work as many or more hours than before, they are free to take an occasional sunny day for play, and make up for it with a few late night or weekend hours—without asking permission or feeling guilty. Some have been so successful that they hired others and built good-sized consulting or other service companies.

Freelance work can be very rewarding for people who thrive on freedom and don't fit well into the structure of corporate life. But it can be hard on others.

DISADVANTAGES

Unless you have guaranteed contracts with more than one company, income can be erratic. And even with contracts, your services can be terminated so quickly and brutally it will make your layoff or early retirement seem kind by comparison.

You are responsible for paying your own office expenses—telephone, travel, printing of brochures and business cards, supplies, computers, secretarial help—and your own taxes. Self-employment taxes are nearly double your former FICA (Social Security) payroll deductions, because now there is no employer matching your contribution. Of course, you will have to pay for your own medical and life insurance coverage, which will cost you more than it cost your former employer. There are state and local business taxes to pay. Further, it is not easy to get credit—either for business or personal reasons—before you've shown two or three profitable years.

Even when finances aren't a factor, many people find that, when the structure and support of the organization is gone, they just can't work as effectively. They need the organization around them to be motivated and productive. On their own, they get little accomplished.

Others miss office camaraderie. Although they spend time with people in the course of providing services to clients, it's not the same as working with a core group of people who remember their birthday and ask about the kids.

If you seek the freedom and income opportunities of self-employed status, evaluate your skills inventory and your personal characteristics to determine if you can:

▪ Motivate yourself;

▪ Work alone;

▪ Make presentations and proposals to win new business;

▪ Provide a living for yourself and your family while meeting business expenses and tax liabilities.

Many people have, and are happier for it.

PART-TIME EMPLOYMENT

You may be financially secure enough to sacrifice some income in exchange for an opportunity to spend more time on hobbies or volunteer work. If you want to remain active in your career on a reduced-schedule basis, part-time employment may be just your cup of tea.

And if that's the case, there are many avenues of opportunity for you. Part-time opportunities have increased in recent years, according to an article in the May 1990 issue of *Changing Times* magazine.

> Part-time employment has grown 21 percent since 1980, according to the National Planning Association. The Bureau of Labor Statistics reports that about 20 percent of employed women and 6 percent of employed men voluntarily work a permanent part-time schedule. And almost 20 percent of part-time jobs are for professionals, which generally means the jobs require a college degree or equivalent experience.
>
> Hundreds of companies are interested in following the lead of firms like US Sprint, which last August began a nationwide program to encourage part-time and job sharing arrangements among its 16,000 employees. "Demographics show we will need flexible work arrangements to get employees with the right skills," says Deb Holt, human resources manager at US Sprint in Kansas City. "Organizations starting to do this now will be in a good competitive position later on."[1]

And, according to the *Changing Times* article, older workers make up one of the fastest growing groups in the part-time work force. However, finding a part-time position can be a job in itself. If you have skills and experience that lend themselves to part-time work, you might have to create your own part-time opportunities.

> [Y]our chances of getting a professional position depend largely on your field and the companies you approach. . . . The best opportunities are in jobs that can be independent or project-oriented. If your experience includes writing or computer programming, for example, those would be attractive skills to emphasize. High stress fields, such as social work, often offer part-time hours to avoid employee burnout. At US Sprint, part-time schedules are available in customer service, field sales, telemarketing, and technical positions. Aetna Life & Casualty is using part-time schedules to help recruit accountants, auditors, and bond raters.[2]

Engineering your own part-time job might involve approaching your current employer and proposing a job-sharing arrangement with a like-minded coworker. If that is not possible, part-time working arrangements are most likely to be found in the following types of organizations:

■ *High-growth, high-tech companies,* because these firms are competing for scarce talent and must be flexible in work arrangements. Recall the discussion in Chapter 1 on industry life cycle, which explained why companies in the early stages of growth offer the best opportunities.

■ *Nonprofit organizations,* such as libraries, museums, universities, and human services organizations, while not the highest paying employers, are often an abundant source of part-time (and rewarding) work.

■ *Small businesses* have a great need for experienced professionals and are less bogged down in bureaucratic hiring policies. As mentioned earlier in the book, nearly 90 percent of workers displaced from large companies find new jobs in smaller, younger companies.

∎ *Federal, state, and local governments* A 1978 law required federal agencies to expand high level part-time positions at salaries and benefits prorated according to number of hours worked. However, professional slots are often restricted to applicants who have worked for the government at least three years.

Many state and city government agencies offer part-time positions, as well.

In most markets in the United States, there are many low pressure (and relatively low-paying) part-time positions available in the service sector: fast food outlets, retail stores, day-care centers, filling stations, and supermarkets. Although most of these jobs traditionally start at the minimum wage level, in places where labor is scarce the hourly rates are much higher.

Getting a salary proportional to a job's full-time equivalent is usually tougher when you join a new company part-time than when you convert a full-time position to part-time. Exceptions to that rule come from a new, and growing, industry: temporary employment for executive and professional positions.

PROFESSIONAL AND EXECUTIVE
TEMPORARY SERVICES

Temporary employment has almost completely shed its clerical and technical image. A full 90 percent of all companies in America use "temps," and placement of executive temporaries is the fastest growing segment of this burgeoning multibillion-dollar-a-year market.

Employment industry insiders have dubbed the new service "head renting" and it's almost exclusively *high* rent. Some executive temporary positions command hourly rates equivalent to a salary of $100,000 a year, and more.

Executive Recruiter News claims there are 30 U.S. firms currently placing executive temporaries, and as many as 200,000 "exec temps" in the field. Job assignments last from three to nine months, and about 75 percent of them involve temporary relocation.

According to the July 15/August 15, 1990 issue of "Special Report," a service of *Executive Recruiter News,*

> Executives placed by "exec temp" firms are pre-screened, high level professionals often earning more than $100,000 a year and no longer saddled with a stigma about temporary work. They are recycled retirees, job seekers (an estimated 70 percent turn temp work into permanent posts), golden parachuters, lifestyle seekers or refugees from the corporate routine, managers-turned-mothers and career interim executives. . . .
>
> They are hired on a project basis, as a fill-in during a search, for their special expertise, or as a tryout. Interim executives may be required by a small company with no succession planning, a company whose corporate downsizing has left it too lean, or by the many top managements now oriented to extremely short-term . . . planning. . . . Exec temps are skilled executives able to transfer extensive knowledge from situation to situation, industry to industry.[3]

If you think executive and professional temporary employment might suit your current career objective, look again at this sentence in the preceding paragraph: *"Exec temps are skilled executives able to transfer extensive knowledge from situation to situation, industry to industry."*

In many cases, "exec temps" are performing work that might have gone to management consulting firms, more evidence of the fluctuating industry life cycle. The article also mentioned that many of the firms requesting temporary executives were in "crisis management," going through bankruptcy or reorganization proceedings.

If you enjoy fighting fires, but get bored sticking around after the fire's out and business is back to normal, this type of temporary executive employment might be a good target for you. The following is a list of "leading" firms in the exec temp field, according to *ERN*:

❚ Siebrand-Wilton Associates, New Jersey and New York

❚ Interim Management Corp., with offices in Los Angeles, New York, and Stamford, Connecticut

❚ Drakewood Resource Co., New York

▪ Alterna-Track (serving financial service firms exclusively), New York

▪ Senior Career Planning & Placement Service, a New York-based division of the National Executive Service Corps. It places volunteer executive consultants—mainly retirees—into nonprofit enterprises to perform project-basis, high-level work.

The following firms place financial and general management executives for temporary assignments:

▪ Accountants on Call, New York City

▪ CFO Associates Inc., Parsippany, New Jersey

▪ David C. Cooper & Associates, Atlanta

▪ Klivans, Becker & Smith, Cleveland

▪ Maximum Management, New York City

▪ Management Assistance Group, West Hartford, Connecticut

▪ Princeton Entrepreneurial Resources, Princeton, New Jersey

▪ Rader, O'Neal, McGuinness & Co, Houston

▪ Robert William James & Associates, Boulder, Colorado

▪ Romac & Associates, Portland, Maine & affiliate offices

▪ Savants, Atlanta

▪ Simpson Nance & Graham, Atlanta

▪ Staff Alternative, Stamford, Connecticut

▪ The Pickwick Group, Wellesley Hills, Massachusetts.

If you're not at the professional or managerial level, check out the oldest and largest temporary agency: Kelly Services (formerly "Kelly Girl"). In addition to temporary assignments in the traditional categories of accounting, clerical, and light industrial jobs, they now place people in marketing, sales, records management, technical support, and home health aid jobs.

VOLUNTEERING FOR FUN AND PROFIT

Nonprofit organizations have been dubbed the "third sector" of the economy and its fastest growing sector, according to an August 6, 1990 *Wall Street Journal* article in which Peter Drucker was quoted as saying that nonprofit entities often "excel where for-profit companies fail [b]ecause they don't focus just on the bottom line, they truly understand the customer."[4] This accelerating growth in the nonprofit sector of the economy may very well coincide with your own desire to pursue a new career path and contribute to an important cause while at the same time earning a living.

An ideal way to "break in" to nonprofit work is to volunteer for a position to gain the experience you need. If you think you'd be good at public relations, youth leadership, health care, or any one of the myriad services required by nonprofit groups, but your career to date has not given you any such experience, contact an organization that has such needs and offer your time and talents free of charge.

The contacts you make and the experience you gain from investing your own time and effort could pay off handsomely in a paid position with that organization or another. It is also a good way to find out if you enjoy doing that kind of work.

Nonprofit organizations that use volunteer services are all around you. The contents of your mailbox on any given day will yield some interesting leads. If you're stuck for ideas, look up the local chapter of the United Way, American Cancer Society, March of Dimes, or American Heart Association in your telephone directory.

The telephone directories of some metropolitan areas (such as Boston) feature a "Human Services Yellow Pages" section, which includes public agencies, nonprofit groups, and fund-raising firms. If your directory does not have such a section, ask your reference librarian for a directory of nonprofit organizations.

SELECT YOUR TARGET—AND HIT IT

Job seekers of all ages deny themselves an opportunity for career fulfillment when they fail to begin a job search with a targeted objective. Their sporadic methods can best be described as "ready, fire, aim." As a result, they don't hit *anything* that remotely resembles a carefully planned career move.

Go through the exercises in this chapter several times. Evaluate thoroughly and honestly. The success of your search —your life, even—hinges on the work you do now in the planning stages. Give it your best. Involve your spouse and/or other family members in the process. Their future happiness may be tied to yours—and vice versa. Without their support, your efforts may lead nowhere. A solid plan will help you achieve that unique and satisfying goal of being able to "spend your life in your own way."

NOTES

1. "Tactics That Win Good Part-Time Jobs," *Changing Times,* May 1990, Volume 44, No. 5, p. 61.

2. Ibid.

3. "Head-Renting. How Hot New Niche Affects Search & Consulting," *Special Report* July 15/August 15, 1990, c. 1990 by *Executive Recruiter News,* Kennedy Publications, Fitzwilliam, New Hampshire.

4. Solomon, Jolie, "Drucker Puts Name on Nonprofit Group," *Wall Street Journal,* Dow Jones & Company, Inc. August 6, 1990.

7 | Researching Your Search

As a rule . . . he (or she) who has the most information will have the greatest success in life.

—Disraeli

Collecting the information you need to land the job you want is the focus of this chapter. I'm going to list enough resources to keep you reading until the next century, but I caution you not to let research stall your search.

Research is such a nonconfrontational, relatively stress-free part of the search process that some job seekers happily hole up in their local libraries and never make a single phone call or send a single resume. (A few have been known to be dusted nightly by the custodian.)

You don't have a lot of time to lose. You're there to locate the resources you need, get the information from them most applicable to your search, then get down to business and *get a job!* And do it all before *your* perfect job is filled by someone who got there ahead of you. Before you hit the local library, consider the following.

GET ORGANIZED

Buy some lined 5 × 7 cards, 5 × 7 dividers with alphabet tabs, and a file box to keep them in. Or, if you prefer, use a three-ring binder stocked with ruled, three-hole-punched paper and alphabetized dividers. This will help you keep all your research notes in one place *and* record the results of your initial telephone and mail contacts. (I prefer the three-hole notebook method; it allows you to keep annual reports, company brochures, and other literature in the same place as your notes.) Figure 7–1 is the form you will use for *each* prospective employer.

Figure 7–1 Prospect Profile/Contact Record

PROSPECT PROFILE/CONTACT RECORD

Company Name: _____

Address: _____

Phone: _____

Contacts:

Name	Title	Phone/Ext
_____	_____	_____
_____	_____	_____
_____	_____	_____
_____	_____	_____

Years in Business: _____

Owned By: _____ Year Acquired: _____

☐ Requested annual report on: _____ Rec'd on: _____

Products/Services of Company: _____

_____

Customer or Client Profile: _____

_____

Figure 7–1 *(Continued)*

Annual Sales/Market Share by Product, Line, or Division: _____
_____
_____
_____

Subsidiaries/Other Locations: _____
_____

Primary Competitors: _____
_____

Problem Areas: _____
_____

Potential Growth Opportunities for Company: _____
_____

Target Positions Within Company: Hiring Official
_____   _____
_____   _____
_____   _____

Date of Initial Contact, Person(s) Contacted and Results:
_____
_____
_____
_____

		Resumes	
Date	To Whom Sent	Date Followed-Up	Interview Date
____	_____	_____	_____
____	_____	_____	_____
____	_____	_____	_____

		Interviews	
Date	With Whom	Thank You Sent	Other Follow-Up
____	_____	_____	_____
____	_____	_____	_____
____	_____	_____	_____
____	_____	_____	

RESEARCH FOR TWO REASONS

Your initial research is designed to help you target your job search—to match your career objectives with the employers most likely to fulfill them. But research is also important for the active part of your search—the campaign—discussed in Part III.

When you do your homework, you can cut through the red tape that hopelessly entangles most searchers. You will have names of decision makers, know how to reach them; and, when you correspond or talk on the telephone, your letters and telephone calls will be better received because you'll sound professional and knowledgable. You will know exactly how your skills and experience can meet the company's needs. In the job search, knowledge *is* power.

HONE YOUR RESEARCH SKILLS

To avoid doing the same work twice, keep your second research objective in mind during the targeting phase. When you encounter a company that appears to be a good prospect, note the address and telephone numbers you will need, take brief notes about the company, call or write for its annual report and/or brochures, then move on to the next prospect.

Gather just enough information for your initial telephone calls and letters, but don't get bogged down recording every detail of each company's operations. The field of companies you eventually interview with will narrow. Prior to each interview, you can go back and look up more details if you need to.

SIC NUMBERS

The United States government has classified every line of business in the country by number. Since many of the resources listed in this chapter use this "Standard Industrial Classification (SIC)" numbering system, it is helpful to understand how it works. Here, from The Transition Team's *Professional Guide to Career Life Planning,* is an example:[1]

> The S.I.C. number consists of four digits:
>
> 0111
>
> The first two digits indicate the type of business:
>
> *01*11 signifies *Agriculture*
>
> The second two digits indicate the specific product or service:
>
> 01*11*
>
> The S.I.C. numbers are constant so that you can look up the S.I.C. number for your particular line of work and be prepared to use it. Here are two more S.I.C. examples:

02 11	35 23
Agricultural Beef Cattle Production Feedlots Livestock	Machinery Farm Machinery

References such as Dun & Bradstreet's *Million Dollar Directory* use S.I.C. numbers to organize the data they contain. For a complete list of S.I.C. numbers, ask at your local library to see a copy of the "S.I.C. Manual."

RESOURCES

Libraries—and librarians—are a treasure trove of information for the serious, self-directed searcher. Librarians (bless them!) love to answer questions. You will be on a first-name basis in no time.

Some public libraries even have special sections—and librarians—devoted to job search materials. The following list contains examples of such libraries around the country and their telephone numbers. Check with your local library, or state library system, to see if they offer the equivalent:

▪▪ *District of Columbia*
  Martin Luther King Memorial Library   (202) 727-1181

∎ *Georgia*
Athens Regional Library   (404) 354-2620

∎ *Idaho*
Idaho Falls Public Library   (208) 529-1462

∎ *Illinois*
Chicago Public Library   (312) 321-3460
Skokie Public Library   (708) 673-7774

∎ *Maryland*
Enoch Pratt Free Library (Baltimore)   (301) 396-5394
Prince George's County Memorial Library (Hyattsville)
     (301) 699-3500

∎ *Michigan*
Detroit Public Library   (313) 833-4251
Flint Public Library   (313) 232-7111
Lansing Public Library   (517) 374-4600
Peter White Public Library (Marquette)   (906) 228-
     9529
Willard Library (Battle Creek)   (616) 968-8166

∎ *Missouri*
St. Louis Public Library   (314) 241-2288 Ext. 358

∎ *Nebraska*
Columbus Public Library   (402) 564-7116
Crete Public Library   (402) 826-3809
Holdrege Public Library System   (308) 995-6556
Slagle Memorial Library (Alliance)   (308) 762-1387
John A. Stahl Library (West Point)   (402) 372-3831

∎ *New Jersey*
Camden County Library (Voorhees)   (609) 772-1636

∎ *New York*
Bethlehem Public Library (Delmar)   (518) 439-9314
Brooklyn Public Library   (718) 780-7777
Buffalo and Erie County Public Library (Buffalo)
     (716) 846-7079
Crandall Library (Glens Falls)   (518) 792-6508
Hempstead Public Library   (516) 292-8920
Johnstown Public Library   (518) 762-8317
Kingston Area Library   (914) 331-1474

Manhasset Public Library   (516) 627-2300
Mid-Hudson Library System (Poughkeepsie)
    (914)471-6060
New York Public Library   (212) 340-0835
Onandaga County Public Library (Syracuse)
    (315)425-5262
Ossining Public Library   (914) 941-2416
Plainview-Old Bethpage Public Library   (516) 938-
    0084
Queens Borough Public Library (Jamaica)
    (718)990-0800
Rochester Public Library   (716) 428-6797
Schenectady County Library   (518) 381-3500
Wadsworth Library (Genesco)   (716) 243-0440
West Islip Public Library   (516) 661-7080
Westchester Library System (Elmsford)   (914) 592-
    8214

▮   *North Carolina*
Durham County Public Library   (919) 683-2626
Forsyth County Public Library (Winston-Salem)
    (919) 727-2680

▮   *Ohio*
Columbus Metropolitan Library   (614) 222-7180
Cuyahoga County Public Library (Maple Heights)
    (216) 475-2225

▮   *Pennsylvania*
The Carnegie Library of Pittsburgh   (412) 622-3133
Chester County Library (Exton)   (215) 363-0884
Citizens Library (Washington)   (412) 222-2400
Monessen Public Library and District Center
    (412)684-4750
Free Library of Philadelphia   (215) 686-5436
Scranton Public Library   (717) 348-3020

▮   *Texas*
Corpus Christi Public Library   (512) 880-7004

▮   *Tennessee*
Memphis-Shelby County Public Library and Informa-
    tion Center   (901) 725-8832

▮▮ *Washington*

Everett Public Library   (206) 252-9053
Timberland Regional Library (Olympia)
    (206)943-5001

Following, from the general to the specific, are titles of national directories and other resources. Your library may even contain their state and local equivalents—for your own area as well as others:

▮▮ *National Directories—General Business*

Larger public libraries will have current copies of many of the following, or may be able to borrow them through their state interlibrary loan system:

Business Periodicals Index (H. W. Wilson Company)
Dictionary of Occupational Titles
Directory of Corporate Affiliations
Directory of Directories
Dun and Bradstreet's Million-Dollar Directory
Dun & Bradstreet's Middle Market Directory
Dun & Bradstreet's Reference Book of Corporate Managements
Encyclopedia of Associations
Everybody's Business
F&S Index of Corporations and Industries (current articles from newspapers and magazines about employers)
Forbes: Annual Report of American Business
Forbes 500
Fortune 500
Guide to Occupational Exploration
International Businessmen's Who's Who
MacRae's Blue Book
Moody's Industrial Manual (and other Moody's manuals)
Moody's News Reports
Occupational Outlook Handbook
Standard Directory of Advertisers
Standard & Poor's Corporation Records

Standard & Poor's Register of Corporations, Directors,
*   and Executives*
Thomas Register of American Manufacturers
Thomas Register of Grocers
U.S. Bureau of Labor Statistics: Area Wage Surveys
U.S. Industrial Product Directory
Value Line Investment Surveys
Wall Street Journal Index
100 Best Companies to Work for in America

▮ *National Directories—Specific Business*

These specific directories contain a wealth of information for the targeted search. You can order from the publisher (addresses in *The Directory of Directories*), but that gets expensive.

Instead of investing hundreds of dollars in directories, ask a local business operating in the targeted industry if you can spend an hour in their offices at their convenience to use their copy. Make your inquiry politely and professionally. If they grant your request, arrive promptly and in business attire; restrict your visit to the allotted time; and conduct yourself as if you are interviewing—you may be!

Accredited Institutions of Post Secondary Education
American Aircraft, Missiles & Satellites Directory
American Apparel Manufacturers' Association Direc-
*   tory*
American Bank Directory
American Druggist Blue Book
American Men and Women of Science
American Register of Importers & Exporters
American Society of Training and Development Di-
*   rectory*
Bankers' Almanac & Yearbook
Book Publishers Directory
Brown's Directory of North American Gas Companies
Canadian Trade Index
Chemical Guide to the United States
Consultants & Consulting Organizations Directory
Davison's Textile Directory

Directory of American Firms Operating in Foreign Countries
Directory of Executive Recruiters
Directory of Franchising Organizations
Directory of Gas Utility Companies
Directory of General Merchandise, Variety & Jr. Dept. Chain Stores
Directory of Oil Well Supply Companies
Directory of United States Importers
Education Directory, State Education Agencies
Executive Directory of the U.S. Pharmaceutical Industry
Foreign Service List
Glass Factory Directory
Guide to Venture Capital Sources
Handbook of Independent Marketing Advertising Services
Hotel Motel Directory/Facilities Guide of the Hotel Sales
Management Association
Jack O'Dwyer Directory (Public Relations)
Machine & Tool Directory
Media Personnel Directory
Medical & Health Information Directory
Metal Manufacturers' Directory
National Association of Personnel Consultant's Directory
National Food Brokers Association Directory
Paper Industry Management Association Directory of Members
Progressive Grocers Marketing Guidebook
Recreation and Outdoor Life Directory
Research Centers Directory
Standard Directory of Advertising Agencies
The International Directory of Computer & Information System Services
Thomas Register of Grocers
Training & Development Organizations Directory
Venture Capital Directory
Wards Automotive Yearbook
Who's Who in Advertising

Who's Who in Electronics
Who's Who in Finance and Industry
Who's Who in Railroading

▮▮ *Trade and Professional Association Resources*

Most professional and trade associations publish their own lists of corporate members and their officers. Even individual members of associations living near you could be valuable contacts when you begin job prospecting. Use the *Encyclopedia of Associations* to get the information you need.

Following, listed by publisher, are trade journals that may yield valuable knowledge about targeted industries and companies. (Many also have an employment classified section.) Although most are sold exclusively on a subscription basis, you can contact the publisher and ask to buy the most recent copy to determine if you ought to subscribe. Many will even send a complimentary copy. If journals for your specific industry or profession are not listed here, you may find them in the *Standard Periodical Directory,* published by Oxbridge Communications.

Published by Cahners Publishing Co.
1350 Touny Ave.
Cahners Plaza
Des Plaines, Illinois 60017-5080
(708) 635-8800

Design News
Purchasing Magazine
Plastics World
EDN
Building Design & Constructions
Building Supply Home Center
Construction Equipment
Foodservice Equipment Specialist
Restaurants & Institutions
Packaging
Professional Builder
Security

Security Distributing & Marketing
Hotels & Restaurants International

Published by Crain Communications, Inc.
740 N. Rush Street
Chicago, IL 60611-2590
(312) 649-5200

Advertising Age
Business Insurance
Pensions & Investments
Rubber & Plastic News
Crain's Chicago Business
Crain's Cleveland Business
Crain's Detroit Business
Crain's New York Business
Automotive News
Modern Healthcare
American Laundry Digest
American Coin-op

Published by Fairchild Publications
55 Fifth Avenue
New York, New York 10003
(212) 741-4000

Electronic News
American Metal Markets
Women's Wear Daily
Daily News Record
Home Fashions Magazine
Supermarket News
Footwear News
Golf Pro Merchandising

Published by McGraw Hill Publications
P.O. Box 900
New York, NY 10020
(212) 391-4570

Aviation Week and Space Technology
Business Week

Coal Week
Chemical Engineering
Electrical Marketing Newsletter
Architectural Record
Coal Age
Data Communications
Electrical Construction & Maintenance
Electrical Wholesaling
Modern Plastics
National Petroleum News
Textile World

THE CLASSIFIEDS

The employment classified ads in metropolitan newspapers are *one* source of job information. According to employment industry estimates, however, only between 15 and 20 percent of available jobs are ever advertised. By the time you see the advertisement, the position may already have been filled by someone from inside the company, or from someone who got there ahead of you through effective networking.

That is why research is so important. It helps you find openings *before* they are advertised. Don't discount classified ads altogether; just don't rely on them solely. Job seekers who use the classifieds as their only resource foreclose 80 percent of their opportunities.

In addition to local and metropolitan newspapers (with Sunday being the day for heaviest advertising), check the *National Business Employment Weekly* at your local library. Published by Dow Jones and Company, it's a weekly compendium of all the employment classifieds that appear of the *The Wall Street Journal.*

THE TELEPHONE TELLS ALL

To the right person, with the right approach. Another way to research prospective employers is on the telephone. It is a more proactive approach, especially suited to those job seekers who *don't* want to spend a lot of time in the library. In *Jeff Allen's Best: Get the Interview,* Allen advised:

> Sales, marketing, customer service, public relations, and personnel departments are the best sources of information. These are staff functions accustomed to telephone inquiries.
>
> Even a receptionist or switchboard operator can be extremely helpful if they're not too busy, since their jobs are often the nerve center through which information flows. Because they regularly deal with the public, they're accustomed to fielding questions. Establish rapport with these individuals, and if they don't have the answers to your questions, they'll readily route you to someone who does.
>
> Here are some questions you can ask to obtain general company information:
>
> 1. Where is the business headquartered?
> 2. Who owns the business?
> 3. How many facilities does the business have?
> 4. What divisions does the business have?
> 5. How many employees does the business have?
> 6. What are the main products or services of the business?
> 7. What markets does the business serve?
> 8. What are the new products or services of the business?
> 9. What are the annual sales of the business?
> 10. How long has the business been in operation?
>
> You can get even more specific, and ask for the department where you'd like to work. You may find yourself talking to someone who will eventually interview you, so be careful to remain professional, polite—and anonymous. Don't talk about yourself. Learn about the employer.
>
> Instead of asking direct questions about job *openings,* make general inquiries about job *opportunities* and the nature of the operation. If your information-gathering call becomes a job-hunting call, you will be transferred to the human resources department where you'll get the standard, "Send a resume and we'll review it."[2]

Although in this book research is part of the "pre-campaign" phase, that doesn't mean you shouldn't act upon any leads you might uncover now. If you find yourself hot on the trail of a job opportunity, proceed directly to Chapter 9, prepare your resume and cover letter, then check out Chapter 10 for tips on telephone prospecting for interviews.

RELOCATION RESOURCES

If it looks like a good career lead is developing, but it means relocating to another part of the country, research the area before you commit to a second interview or entertain an offer. Use the information you gather from the resources listed here to compare with your earlier evaluation (Chapter 6) of the area where you now live. You'll be able to decide objectively whether a move is indicated.

▪▪ *Places Rated Almanac: Your Guide to Finding the Best Places to Live in America,* by Richard Boyer and David Savageau (Englewood Cliffs, NJ: Prentice-Hall).

▪▪ *The Book of America,* by Neal R. Peirce and Jerry Hagstrom (New York: W.W. Norton).

▪▪ *The Book of American City Ranking* by John T. Martin and James S. Avery (New York: Facts on File Publications).

▪▪ *Moving: A Guide to Selecting a School System,* by Albert and Marilyn Paulter (Williamsville, NY: Paulter Associates).

▪▪ *The Best Towns in America, a Where to Go Guide for a Better Life,* by Hugh Bayless (Boston: Houghton-Mifflin).

▪▪ *The Nine Nations of North America,* by Joel Garreau (Boston: Houghton-Mifflin).

▪▪ *Safe Places (East and West),* by David and Holly Franke (New York: Warner Paperbacks).

▪▪ *Where the Jobs Are,* by William J. McBurney, Jr. (New York: Chilton Books).

▪▪ *Where You Live May Be Hazardous to Your Health,* by Robert A. Shakman, M.C., M.P.H. (New York: Stein & Day).

- ▮ *The World Almanac & Book of Facts.* (New York: Newspaper Enterprise Association).
- ▮ *Moving to.* . . . (Snyder, NY: Moving Publications, Ltd.).
- ▮ *The New Book of American Rankings* (New York: Facts on File Publications).

The previous list was taken from *The Transition Team's Guide to Relocation* by Gail A. Ryder, who recommended the following:

Another excellent source of information about any town is the Chamber of Commerce. They publish a wide variety of brochures and directories that could be very useful in your decision-making process. Their membership directories will give you the names, addresses, and phone numbers of top executives at many of the local businesses. Your reference library should have copies of both the following publications:

- ▮ *World Wide Chamber of Commerce Directory*
  Johnson Publishing Company
  Eighth & Van Buren
  Loveland, CO 80537
- ▮ *Directory of State Chambers of Commerce and Associations of Commerce and Industry Executives*
  Chamber of Commerce of the United States
  1615 H. St., NW
  Washington, DC 20062

Although your local library probably maintains a collection of telephone directories, you should also make a point of obtaining the *Yellow Pages* and other phone directories from areas in which you are interested. Your local telephone company business office has them on file or can get them for you.

A subscription to the local newspaper can reveal a great deal more than what job openings are being advertised. Remember that only about 15 percent of all openings end up in the classifieds, so don't read only the help wanted section. there. Instead, briefly scan each page for clues as to lifestyle, activities, services, and so on.

Many cities as well as smaller towns also have their own full-color magazines, which are published as often as weekly.

Check with "relocation services" sponsored by local boards of realtors. In larger cities, the packages of information they'll prepare will need to be delivered on a flat bed truck! Just remember that this source of information will only accent the positive. Use the worksheet in Figure 7–2 to evaluate the new area.

Figure 7–2 Relocation Checklist

RELOCATION CHECKLIST

City: _____ State: _____

Nearest Large Metropolitan Area: _____

Population:

Climate:
    Average Rainfall:
    Average Temperature Winter:
    Average Temperature Summer:
    Annual Average Temperature:
    Other Notes:

Geography:

Economy:
    Major Employers:
    Types of Industry:
    Unemployment Rate:
    Median Household Income:
    Average Household Income:
    Per Capita income:

Education:
    Public School System Size:
    Ranking:
    Test Scores:
    Private Schools
    Colleges/Universities:

Figure 7–2 *(Continued)*

Housing:
    Median Home Price:
    Average Price for Comparable Home:
    Average Number of Listings on Market Past 6 Months:

Health Care and Environment:
    Air Quality:
    Water Quality:
    Toxic Problems:
    Access to Health Care:

Recreation:

The Arts:

Transportation:
    Distance to Major Airport:
    Public Ground Transportation:
    Rail Services:

Crime and Law Enforcement:
    Safer?
    Less safe?

Additional Notes:

With your resources in hand and your research nearly complete, you're ready to launch your job campaign. Part III will tell you everything you need to know to write powerful resumes and cover letters, direct them to the decision makers who will be most interested in them, activate your network of job leads and references, and interview successfully.

But before you get down to the business of getting that job there's some information on age discrimination I think you should know. Forewarned is forearmed. Chapter 8 gives you the details to help you detect—and defeat—age discrimination in hiring and on the job.

NOTES

1. *Professional Guide to Career Life Planning,* Copyright 1989 The Transition Team Network, Inc., All Rights Reserved. The Transition Team Network, 3155 W. Big Beaver Road, Suite 117A, Troy, Michigan.

2. Allen, Jeffrey G., J.D., C.P.C., *Jeff Allen's Best: Get the Interview,* New York: John Wiley, 1990, pp. 10–11.

3. Ryder, Gail A. *The Transition Team's Guide to Relocation,* Copyright 1986 The Transition Team Network, Inc. All Rights Reserved. The Transition Team Network, 3155 W. Big Beaver Road, Suite 117A, Troy, Michigan.

PART III | The Campaign

8 | Defeating Age Discrimination

In preparing for your job search after 50, you must be aware of the opportunities for—*and* the obstacles to—fulfilling employment. Much of the book focuses on opportunities. In this chapter, I discuss one of the major pitfalls you may encounter—age discrimination—and how to recognize and overcome it.

Age discrimination is illegal. The federal government, as well as many state and local governments, have laws against it. Although age discrimination is still prevalent in the work place, workers are becoming more aware of their rights—and enforcing them.

DISCRIMINATION COMPLAINTS INCREASE

In 1986, older workers filed more than 17,100 charges of age discrimination with the federal government's Equal Employment Opportunity Commission (EEOC). The charges covered alleged abuses in hiring, promotions, and compensation policies. That number of complaints was almost twice the number filed in 1981, evidence that older workers are becoming more aware of their rights and recourse in matters of age discrimination.[1]

AGE DISCRIMINATION IN EMPLOYMENT ACT

Recognizing that age discrimination in the work place indeed exists and creates hardship—even crisis—for its victims, the U.S. government in 1967 passed the Age Discrimination in Employment Act (ADEA) to combat age discrimination and provide remedies for it. The ADEA was created to:

▮▮ "Promote employment of older persons based on their ability rather than age;

▮▮ "Prohibit arbitrary age discrimination in employment;

▮▮ "Help employers and workers find ways of meeting problems arising from the impact of age on employment."[2]

The ADEA prohibits discrimination in employment against most workers age 40 and older. It is administered through the EEOC, which has offices nationwide. Addresses and phone numbers for EEOC offices are listed at the end of this chapter.

Under the ADEA, employers with 20 or more employees may not:

▮▮ Discriminate against workers age 40 and older in hiring, discharge, or any other aspects of employment because of age;

▮▮ Indicate any preference based on age in notices or advertisements for employment;

▮▮ Take action against any employee for complaining about age discrimination or for helping the government investigate an alleged case of age discrimination.

Also, employment agencies serving one or more covered employers may not discriminate against workers age 40 and older in referral because of age.

EXCEPTIONS TO THE RULES

There are exceptions to the ADEA, which employers can use as legal defenses against allegations of age discrimination. Under the ADEA it is legal to:

▪ Take an action based on age where the use of age as a basis for employment is necessary to the normal operation of the business, for example, hiring a young actor to play a child's part;

▪ Observe the terms of a bona fide seniority system or any bona fide employee benefit plan;

▪ Take an action based on reasonable factors other than age;

▪ Discharge or otherwise discipline an employee for good cause;

▪ Retire certain executives or high policymaking employees at age 65;

▪ Retire tenured faculty at institutions of higher learning at age 70, effective January 1, 1987 through December 31, 1993;

▪ Discharge or refuse to hire firefighters or law enforcement officers under applicable state and local laws, effective January 1, 1987 through December 31, 1993.[3]

Aside from the listed exceptions, the ADEA prohibits almost all employment actions in which age has been used as a factor, even when there are other legitimate reasons for the decision. It also prohibits discrimination within the protected group, for example, a preference on the basis of age for a 45-year-old employee over a 50-year-old employee.

Most states, and even many local governments, also have laws against age discrimination. If state or local laws provide greater protection than the federal law, the state or local law takes precedence.

In spite of more than 20 years of legislation against it, age discrimination still occurs in hiring, firing, and all employer-employee transactions in between. Recognizing it is the first step to overcoming it.

DISCRIMINATION IN HIRING

Age discrimination in hiring is often the most difficult to detect. Those "thanks but no thanks" form letters that come in the mail don't say, "We think you're too old for the position." For example:

> George F., 52, a structural engineer, sent his resume to a commercial construction company in reply to a classified advertisement of job openings. He received a letter from the company expressing interest in his qualifications, along with an application to be completed and returned "as a formality" before an interview would be scheduled.
>
> George filled out the application, which included a detailed work history and dates, and returned it. Instead of scheduling an interview, the company sent George another letter which stated that it had no openings suited to George's qualifications.

Did George's qualifications change in the space of a week? No. More likely, although it's difficult to prove, the hiring official did some quick calculations based on George's work history and determined George was "too old." At other times, the discrimination is more obvious.

> Susan T., 41, had excellent experience and qualifications in banking and finance. When she interviewed for the position of loan officer with a large commercial bank, she was told that she wouldn't be considered because the bank's "profile for these positions was a recent college graduate."

In other words, management had a policy and a practice of hiring only young people for certain positions. In this case, even someone in his or her thirties would be discriminated against as "too old" for the job.

OVERCOMING DISCRIMINATION IN HIRING

What could George have done? How should Susan have responded? Discrimination in hiring is difficult to detect and prove. Also, it is time consuming and costly to pursue a cause when all you set out to do was further your career. And, since you don't yet *have* the job, it is tough to prove damages from *losing* it. Besides, you need to direct your time, energy, and positive attitude to your job search. A fight can drag you down right now.

My advice to George would be to call or write the personnel officer who had signed the letter and say, without malice or anger: "Since my professional qualifications did not change from our first correspondence to our second, I can only assume that some personal information revealed on my application caused you to disqualify me from consideration. I am still very much interested in the position, believe I would be an asset to your company, and would appreciate the opportunity to present my qualifications more fully in an interview."

As for Susan's interview encounter, I would have suggested she smile and say, "I believe that a practice of hiring only workers of a certain age for a job is illegal. I urge you to consider that my experience and discretion would be an asset in the important task of approving large commercial loans."

Then I would tell them both that the only result they should expect is the satisfaction of having spoken their minds.

Discrimination in hiring persists because of the stereotypical notions described in earlier chapters. Sometimes it is so ingrained that company policy makers and hiring officials don't even recognize that their practices are discriminatory. Does the job seeker benefit from pointing it out?

No. I know of *no* case in which an applicant alleged discrimination and got the job.

The bank vice-president who told Susan she didn't "fit the profile" didn't even stop to consider that such a policy was discriminatory. Susan's informing him of that fact won't win her the job, either. In addition, he may well want to avoid someone obviously well-versed in her rights.

The best way to fight discrimination *before* you're hired is to avoid it. And the best way to do that is to get to the interview stage without ever revealing your age. Chapter 9 gives you tips for writing a resume that highlights your accomplishments and skills, not your age. Chapter 11 describes ways to overcome age prejudice in an interview.

In Chapter 4, "Take Ten Years Off Your Image" I gave you a rule that warned, "Don't Apologize for Your Age." While I'm not breaking the rule, I'm bending it a little here. If you think you're up against age prejudice in an interview situation, and you want to keep yourself in consideration for the job, you *can,* under those circumstances, say something like . . .

> I sense that you may be concerned about my age. I believe that my age is an asset in this position, not a liability. Not only have I continued to learn and grow throughout my professional life, keeping my skills and knowledge up to date, but I have the seasoning and judgment this job calls for. Even if I *could* turn back the clock ten years, I wouldn't. I wouldn't be as qualified for this job ten years ago as I am today. I'm sure you're aware of the studies that prove older workers (executives, managers, etc.) are more loyal and hardworking than younger workers while remaining as productive as younger workers, *aren't you?*

By ending your "speech" with a question (and a friendly smile), you can gracefully step down from your soapbox and give the interviewer the opportunity to agree with you.

> Marvin T., age 62, encountered a similar situation. An international negotiator and an accomplished triathlete, Marv was asked (by a much younger interviewer) "How old are you?"
> Marv replied, "I think what you're asking me is 'How long will I be in this position?'" Then Marv paused and said firmly, in a style that demonstrated his keen negotiating skills, "I'm going to be here at least five years. How many younger candidates can promise you that?"
> Marv got the job.

Here are some other age-related questions you may be asked:

■ How would you feel about working with younger colleagues?

■ How old are your children?

■ Do you have grandchildren?

These questions crop up more often in interviews outside the human resources department. Because personnel people are aware of discriminatory practices, they tend to avoid them.

Such questions are not illegal *per se.* If you were to file a complaint with the Equal Employment Opportunity

Commission, you would have to prove that the question was asked for the purpose of discriminating against you.

The best way to answer these questions is to be prepared for them and not get caught off guard. Your answer will depend on how badly you want the job. If you decide that the questions are too personal and you wouldn't take the job for any amount of money, refuse to answer and end the interview. You might be doing the next applicant a favor. Or, you could lightheartedly counter with, "I'm 50—and twice as good as I was at 25!"

Then there's the sincere approach: "I am 55. I'm in excellent health, and, as you can see from my resume, I have an impressive record of experience and achievement to contribute. I plan to be making a contribution for a good, long time."

Finally, you can redirect the question: "What I would really like to talk about is the success I had in developing new customers for the ABC Company."

Getting past discrimination in the hiring process is only the first step. You're not always home free. There may be many age discrimination traps waiting for you in the work place. That is why I've included here a discussion of discrimination on the job, and how you can counter it.

DISCRIMINATION IN LAYOFFS

Mary H., 55, had worked for a national department store chain as a salesperson for 20 years. Her performance ratings were consistently high. In 1988, she received a regional sales award. As a senior member of the sales staff, Mary earned more than any other nonmanagement employee. In 1989, during a period of "cutbacks," Mary was laid off. At the same time, the store was hiring part-time sales personnel at lower rates of pay.

Laws require layoffs to be based on seniority or merit—the least senior or the least productive employees are the first to be laid off. In this case, that did not occur. Mary was one of the most senior and most productive members of the store's sales staff. By considering Mary's age and salary level as reasons for layoff, the department store violated the ADEA. Mary sued and won reinstatement and back pay.

Millions of dollars in liability have been paid by employers who have attempted to lay off older workers first when reducing numbers of workers. Cases that have been settled in and out of court in favor of employees include:

∎ A life insurance company that laid off 150 older sales representatives during a cutback;

∎ A research facility that laid off hundreds of engineers while recruiting on college campuses for replacements;

∎ A railroad that decided to lay off every worker eligible for a pension rather than younger workers—even though the savings from that action was almost negligible.

In each of the above cases, lawsuits by both the Federal government and the workers resulted in large awards to the workers.[4]

DISCRIMINATION IN DISCHARGES

While factors other than age may contribute to a discharge, if investigation reveals that only older workers were held up to certain standards, while younger workers who showed the same characteristics were kept on the job, then age discrimination has occurred. For example:

∎ An employer may allege that an older employee was discharged because the employee failed to pass a company physical examination. That same employer, however, may have several younger employees remaining on the payroll who also failed to pass the physical—some with even more serious conditions than the older worker.

∎ The reason given for discharging an older employee may be that the employee was given only a "fair" evaluation, and in attempting to get rid of "deadwood" the employer is discharging mediocre employees. However, several younger employees in the same department also received only "fair" evaluations, yet they were retained.

AGE HARASSMENT

It is also illegal to pressure older workers into quitting voluntarily so that they may be replaced by younger workers. It is an employer's responsibility to ensure that this does not take place. If it does, the employer can be charged with age discrimination. For example:

> Frank N., the oldest foreman in his plant, was beyond the company's customary retirement age. Younger workers frequently asked him why he hadn't retired and within his earshot make comments such as, "The old man should retire and give someone else a chance at the job."

What happened to Frank constituted age harassment. Increasingly, older workers are taking action to bring it to a halt.

DISCRIMINATION IN EMPLOYEE BENEFITS

It *is* legal for some employee benefit plans to provide lower benefits for older workers than for younger workers. However, this does not mean that people protected by the ADEA are not entitled to participate in an employer's benefit plans.

Generally, the rule applies to the employer's *cost* of providing benefits. For example, if the same premium payment for life insurance provides lower death benefits for older workers because of the higher risk of this group, that is legal.

Older workers—even those age 65 and older—are entitled to the same medical and pension benefits provided to younger workers.

FILING AN AGE DISCRIMINATION
IN EMPLOYMENT COMPLAINT

To prevail in an age discrimination suit, you must be able to prove that:

▪ You are in the age group protected by the ADEA.

▪ The defendant (employer) in the case was subject to ADEA provisions.

▪ You were adversely affected by some employment action.

▪ Facts indicate that age was a consideration in the action.

If, in your judgment, your case meets all four criteria and you wish to take action, then the following steps should be followed:

1. File a complaint with your employer and, if applicable, a grievance with your union. To file a complaint with your employer, follow the company's internal personnel procedures. Check with your company's human resources department to be sure you are in compliance with any established procedures.

This may be all you need to do to redress your grievance. Upon seeing a complaint of age discrimination, those responsible for the discriminatory action may reverse their decision. Even if they don't, the ADEA still requires you to "exhaust your remedies" before you can qualify for protection by the ADEA.

2. Providing the company's management hasn't come to its senses and ceased its discriminatory practices, the next step is to check with the nearest Equal Employment Opportunity Commission (EEOC) office (addresses at end of chapter)as well as the state government agency that handles age discrimination complaints in your state.

If the rules and procedures of the state law seem more favorable to you than the ADEA provisions, or if the state agency indicates it will prosecute your case faster or more aggressively than the EEOC, file your complaint with the state agency shortly before you file a second one with the EEOC. Usually, the agency that receives the complaint first will handle the case; the other agency will wait to see the result of the first investigation.

Your best bet is to meet with an official in each office in person, if possible. Your case becomes more real than it will over the telephone and you're not as likely to get shuffled back and forth.

3. Even if there is a clear decision for which agency will handle your complaint, you must file a second complaint with the other agency in order to protect your right to sue in court later on. Provisions of the ADEA require that, before you can bring suit under it, complaints be filed with both agencies.

If you file with the state agency first, you must also file a "charge" with the local office of the EEOC. Ordinarily, a charge with the EEOC must be filed within 180 days of the date of the alleged act of discrimination. The days are counted from the date you received the notice from the employer of an action such as a refusal to hire or promote, layoff, or termination. If there is no specific date, as in the case of discrimination that takes place over a period of time, file your charge as soon as you have gathered the information that convinces you that you have been the victim of age discrimination.

Your "charge" must be in writing. State the facts clearly and simply in a letter addressed to the nearest EEOC office. Include in the letter:

∎ Your name and address;

∎ Your employer's name and address;

∎ Your job title, a brief job description, and dates of employment;

∎ Depending on your individual circumstance, a description of the job or work condition that was denied you; or the date you were laid off or terminated and the employer's stated reason for such action; or a description of an ongoing action (such as age harassment) that you believe is discriminatory and your reasons.

Keep the letter brief and to the point. You will be given an opportunity to state your case more fully after the initial charge has been filed.

4. You will receive a written reply telling you to schedule an interview with an investigator who will ask you questions, look at any written evidence you have, and discuss what remedy you seek. Take to that meeting *copies* of any written documents, including company policy manuals and/or employee handbooks, letters, memos, and the like, that you think will help prove your case. (Be sure you keep the originals.)

5. The agency will investigate the charge, and, if it finds age discrimination, attempt to negotiate a settlement. If unable to reach a settlement on your behalf, the agency may file a lawsuit. However, this is a rare occurrence. It is more likely that you will, if you choose to pursue the case, file your own lawsuit through a private attorney or public legal assistance agency.

6. If and when you get to this stage of the process, be sure to select a lawyer (1) who has experience in age discrimination cases, (2) is genuinely interested in your case, and (3) believes you have a chance. Get an estimate of the time and costs up front. If the lawyer is willing to take your case "on a contingency," there's a good chance she or he believes you will prevail. (However, it might also mean she or he doesn't have a lot of other work.) If you're paying by the hour, set a limit on how much you are willing to pay and what you expect as a result of your investment.

CAUTIONS AND PRECAUTIONS

The EEOC has come under fire for delays and mismanagement of age discrimination complaints. According to an article in the May 1988 issue of *50 Plus* magazine, a 1987 Congressional investigation determined that the EEOC let the statute of limitations expire on more than 900 age discrimination complaints. By "dropping the ball," the EEOC made it impossible for complainants to file lawsuits.

Like most government agencies, EEOC is understaffed, overburdened, and outgunned by the legal staffs of most corporate defendants. As a result, its track record leaves much to be desired.

If you file a lawsuit through a private attorney, consider the time, effort, and dollars involved; and weigh all these to determine if they would be better spent in a more positive way. While I agree it is important to "make a point" and stand up for principles, it is also critical to get on with your life.

Don't pin your hopes on a government agency *or* a private attorney to redress wrongs and get your job back. There are new opportunities waiting for you. Isn't your future more compelling than the past?

Now that you have reviewed the past (your skills inventory), targeted the right job for you, and researched where to find that job, let's get on with the important business of writing resumes, getting interviews, and turning interviews into offers. All that is covered in Part III.

APPENDIX

EEOC FIELD OFFICES

Albuquerque EEOC
505 Marquette NW. Suite 1105
Albuquerque, NM 87102
(505) 766-2061

Atlanta EEOC
75 Piedmont Ave., NE Suite 1100
Atlanta, GA 30335
(404) 331-6091

Baltimore EEOC
109 Market Place, Suite 4000
Baltimore, MD 21202
(301) 963-3932

Birmingham EEOC
1829 1st Ave.
N. Heritage Place Building, 5th Floor
Birmingham, AL 35203
(205) 731-0082

Boston EEOC
John F. Kennedy Federal Building
Room 409B
Boston, MA 02203
(617) 565-3200

Buffalo EEOC
28 Church St., Room 301
Buffalo, NY 14202
(716) 846-4441

Charlotte EEOC
5500 Central Ave.
Charlotte, NC 28212
(704) 567-7100

Chicago EEOC
536 So. Clark St., Room 930-A
Chicago, IL 60605
(312) 353-2713

Cincinnati EEOC
550 Main St., Room 7015
Cincinnati, OH
(513) 684-2851

Cleveland EEOC
1375 Euclid Ave., Room 600
Cleveland, OH 44115
(216) 522-2001

Dallas EEOC
8303 Elmbrook Dr., 2nd Fl.
Dallas, TX 75247
(214) 767-7015

Denver EEOC
1845 Sherman St., 2nd Fl.
Denver, CO 80203
(303) 866-1300

Detroit EEOC
477 Michigan Ave., Rm. 1540
Detroit, MI 48226
(313) 226-7636

El Paso EEOC
700 East San Antonio St.
Room B-406
El Paso, TX 79901
(915) 534-6550

Fresno EEOC
1313 P St., Suite 103
Fresno, CA 93721
(209) 487-5793

Greensboro EEOC
324 W. Market St., Rm. B-27
Greensboro, NC 27401
(919) 333-5174

Greenville EEOC
300 E. Washington St.
Federal Building, Suite B41
Greenville, SC 29601
(803) 233-1791

Houston EEOC
1919 Smith St., 6th Floor
Houston, TX 77002
(713) 653-3320

Indianapolis EEOC
46 E. Ohio St., Rm. 456
Indianapolis, IN 46204
(317) 226-7212

Jackson EEOC
100 W. Capitol St., Suite 721
Jackson, MS 39269
(601) 965-4537

Kansas City EEOC
911 Walnut, 10th Floor
Kansas City, MO 64106
(816) 426-5773

Little Rock EEOC
320 W. Capitol Ave., Suite 621
Little Rock, AR 72201
(501) 378-5060

Los Angeles EEOC
3660 Wilshire Blvd., 5th Floor
Los Angeles, CA 90010
(213) 251-7278

Louisville EEOC
601 W. Broadway, Rm. 613
Louisville, KY 40202
(502) 582-6082

Memphis EEOC
1407 Union Ave., Suite 502
Memphis, TN 38104
(901) 521-2617

Miami EEOC
1 Northeast First St., 6th Fl.
Miami, FL 33132
(305) 536-4491

Milwaukee EEOC
310 W. Wisconsin Ave., Suite 800
Milwaukee, WI 53203
(414) 291-1111

Minneapolis EEOC
220 Second St. S., Rm. 108
Minneapolis, MN 55401
(612) 370-3330

Nashville EEOC
404 James Robertson Pkwy.
Suite 1100
Nashville, TN 37219
(615) 736-5820

Newark EEOC
60 Park Place, Rm. 301
Newark, NJ 07012
(201) 645-6383

New Orleans EEOC
701 Loyola, Suite 600
New Orleans, LA 70113
(504) 589-2329

New York EEOC
90 Church St., Rm. 1501
New York, NY 10007
(212) 264-7161

Norfolk EEOC
252 Monticello Ave.
Norfolk, VA 23510
(804) 441-3470

Oakland EEOC
1333 Broadway, Rm. 430
Oakland, CA 94612
(415) 273-7588

Oklahoma City EEOC
531 Couch Dr., 1st Fl.
Oklahoma City, OK 73102
(405) 231-4911

Philadelphia EEOC
1421 Cherry St., 10th Fl.
Philadelphia, PA 19102
(215) 597-7784

Phoenix EEOC
4520 N. Central Ave., Suite 300
Phoenix, AZ 85012
(602) 261-3882

Pittsburgh EEOC
1000 Liberty Ave., Rm. 2038A
Pittsburgh, PA 15222
(412) 644-3444

Raleigh EEOC
127 W. Hargett St., Rm. 500
Raleigh, NC 27601
(919) 856-4064

Richmond EEOC
400 N. 8th St., Rm. 7026
Richmond, VA 23240
(804) 771-2692

San Antonio EEOC
5410 Fredericksburg Rd., Suite 200
San Antonio, TX 78229
(512) 229-4810

San Diego EEOC
880 Front St., Rm. 4S21
San Diego, CA 92188
(619) 557-6288

San Francisco EEOC
901 Market St., Suite 500
San Francisco, CA 94103
(415) 995-5049

San Jose EEOC
280 S. First St., Rm. 4150
San Jose, CA 95113
(408) 291-7352

Seattle EEOC
1321 2nd Ave., 7th Fl.
Seattle, WA 98101
(206) 442-0968

St. Louis EEOC
625 N. Euclid St., 5th Fl.
St. Louis, MO 63108
(314) 425-6585

Tampa EEOC
700 Twiggs St., Rm. 302
Tampa, FL 33602
(813) 228-2310

Washington EEOC
1400 L. St. NW, Suite 200
Washington, DC 20005
(202) 653-6197

NOTES

1. Source: "ADEA Guidebook" copyright 1987 by the American Association of Retired Persons, Worker Equity Department, 1909 K St. NW, Washington DC 20049
2. Ibid.
3. Ibid.
4. Ibid.
5. Ibid.

9 | Writing the Resume

This "paper profile" presents you before you make a personal appearance. In fact, it determines whether you will get the opportunity to appear in person for an interview. Good resumes are powerful advertising tools. Bad ones (and they are plentiful) eliminate a prospect quickly.

Job seekers of all ages and career stages must follow accepted conventions for writing resumes and cover letters. It's a game and you have to know the rules. In addition, 50-plus candidates must know the special rules written just for them so that they don't get "written off" because of age.

THE PRELIMINARIES

Before you gather your exercise sheets and notes from Part II and get ready to write, there are a few details to attend to.

STATIONERY

You'll need quality, engraved stationery for both resumes and cover letters. Stationers and office supply stores are the best sources for high quality stationery. Place your order now so that it will be ready by the time you've finished writing your resume.

The $8\frac{1}{2} \times 11$ inch size is fine for both resumes and letters, but you can also invest in "monarch" (5×7) size for cover letters if you wish. The stock should be white (ivory is acceptable, but white is preferred), at least 24-pound weight. Your name, address, and contact telephone number should be engraved at the top in a traditional serif typeface (Times Roman, Century Schoolbook, for example), not italic. This is business stationery, and it should look businesslike. Order a supply of unengraved second sheets and some matching No. 10 envelopes with return address engraved at the upper left.

Getting a job is a numbers game, and those who win it make dozens, even hundreds of contacts. So, order an ample supply of stationery. Photocopied resumes are simply not appropriate for someone of your age, wage, and career stage.

TELEPHONE TIPS

On your stationery, just below your address, you'll be listing a telephone number. What's it going to be? Think about it.

If you're employed, you have many reasons for keeping your search confidential. Unless you can take calls in complete confidence and privacy and are sure that, in your absence, accurate messages will be taken by someone you trust, don't list your work number.

Listing your home phone number can be a problem, too. Busy signals and no answers will rarely rate a repeat attempt. Family members, no matter how earnest, often sound unprofessional.

Consider installing a separate phone line in your home, with an electronic answering machine hooked up to it. It's a worthwhile investment—and only a temporary one. Get the kind of answering device that allows you to phone in for your messages throughout the day, and you'll be able to return calls while the potential employer is still interested.

Warn all family members away from *your* phone—it's your career lifeline. Don't have the telephone number published in the directory. That way, your friends won't call you on that line, tying it up when you need it.

If you *must* use the family phone:

1. Persuade family members to keep all calls short, especially from 9 to 5; and
2. Train them to take your business messages.

With the important preliminary details out of the way, you can begin in earnest to prepare a powerful, persuasive, interview-winning resume.

RESUMES: WHAT THEY ARE; WHAT THEY ARE NOT

The purpose of a resume is frequently misunderstood. Some of the more common misconceptions:

∎ The resume will get me a job.
∎ The resume must be a complete history of my life and work.
∎ It is content, not form, that counts.
∎ The length of the resume is not important.
∎ People over the age of 50 should use a functional resume format to hide their age.

Let's set the record straight.

THE RESUME WILL GET ME A JOB

False. Only *you* will get you a job. The resume, however, will open the doors to job opportunities. It's not a job-getting tool, but an interview-getting tool.

THE RESUME MUST BE A COMPLETE HISTORY OF MY LIFE AND WORK

False. The resume is a *preview* of the coming attraction, not a feature-length film. Its contents should be carefully designed to make the reader want *more*. Persuasive copy—just enough to arouse curiosity and an interview invitation—is the key to good resume writing.

IT IS CONTENT, NOT FORM, THAT COUNTS

True and False. Both content and form are equally important. Eye appeal is "buy" appeal. An attractive, readable format—with margins, white space, and all information set forth in a way that highlights important points and makes the resume a quick read—will help you sell your talents.

A winning resume is error free. Impeccable spelling and grammar are imperative. Smudges, erasures, or cross-outs are not acceptable. Like your interview suit, your resume represents your image. Fail to follow good resume form, and you won't get a chance to wear the suit.

THE LENGTH OF THE RESUME IS NOT IMPORTANT

False. The resume-length rule is this: Be brief, be brilliant, be interviewed. A resume should be no more than two pages-one, if possible. With resumes, less is more.

Some of the books available today, especially those targeted to "executive" level jobs, contradict the one-to-two page wisdom. The authors of these books advise readers (many of whom are over 50 and in upper management levels) to write three-, four-, even five-page resumes. They base their advice on the direct-mail marketing dictum that "long copy" sells.

Insurance, perhaps. Job candidates, rarely. I've reviewed the results of this advice—I wasn't sure I was reading a resume or a research project.

Some rules of direct-mail marketing apply to resumes, but not all. The job-getting game is unique. Reliable, well-known surveys have shown that the average resume gets 20 to 30 seconds of attention. If you don't "grab and hold" in that brief period, all your efforts are wasted.

That's why you must "wow 'em" with no more than one page—two pages maximum. Even with a two-page resume, the first page should captivate.

Allow me to justify my strong stand. If you are responding to an available position, advertised or not, you are obviously not alone. Someone is going to be reading dozens, hundreds, maybe thousands, of resumes. Standing out in the crowd means being brilliant—in brief form. A one-pound autobiography will keep going to the bottom of the pile until someone has time to read it. That time will never come, and neither will the interview.

If you are sending an unsolicited resume to a decision maker whose name surfaced through your networking efforts, the same rule governs. The typical response of someone who's just received a four- or five-page epic from an unknown source: "Who is this person haranguing me with his (her) life history, and why have I been singled out for punishment?" Not a good first impression.

Finally, if you've discovered someone willing to give you that much time, you're better off skipping the resume step altogether and proceeding straight to the interview. You can bring along supporting documents, if you like.

The resume is your first visual contact. In most cases, putting too much information in it could work against you. (Don't filmmakers put only their best scenes in the previews?) Use it to get past the first checkpoint. There will be time to fill in details later.

PEOPLE OVER THE AGE OF 50 SHOULD USE A FUNCTIONAL RESUME FORMAT THAT HIDES THEIR AGE

False and True. People over the age of 50 *should* avoid giving clues to their age in their resumes, but a functional resume format is decidedly *not* the way to do it.

There are two main resume formats, and several variations on each. The preferred format, even for job seekers with long histories, is the chronological format, which is actually an *inverted* chronological presentation of career achievements, beginning with the most recent.

The functional format groups accomplishments by category (such as "management" "sales"), and omits dates and places. Resume screeners then have to try to piece the puzzle together to figure out when a candidate worked where and for how long. Few bother.

Functional formats were frequently recommended for older job seekers and those who "had something to hide." It didn't take long for the whole world to catch on. If you use a functional format, you'll look like the only person who *didn't* catch on!

RAY'S RULES FOR TAKING YEARS OFF YOUR RESUME

In Chapter 4, I discussed how to take years off your interview image. A few simple techniques do the same for your resume.

DON'T LIST EVERYTHING

Your most recent 15 years of career achievement are more relevant than your earlier work. List only the most recent and highest level positions. You're not "lying by omission," you're simply keeping the resume relevant and brief. Resume writers of all ages should follow this rule.

ELIMINATE OLD DATES FROM EDUCATIONAL CREDENTIALS

The institution attended and degree awarded are sufficient. You needn't say you earned the degree in 1960. If you completed a degree during the past 5 or 10 years, however, it becomes an advantage to include the date. It shows you're staying on top of your career and your accomplishments are current.

LEAVE YOUR FAMILY AT HOME

They're your life, but they don't belong in your resume. Save the space for career accomplishments. "Married, two children ages 17 and 22," is a mid-life giveaway. Personal questions may come up in your interview—unless you give the answers in your resume first. Then, you might not even be asked to show up for the interview.

Having a grown or almost-grown family *should* be viewed as a plus by employers. It usually means fewer demands on an employee's time and lower benefit costs. (No sick children to run home to, no day care problems, no employer-paid maternity costs or orthodontics, to list just a few.) You can make these points in your interview when the question of age comes up. There is no graceful way to say it in a resume, so avoid trying.

Those are the ground rules. Let's discuss the specifics of resume style and content and take a look at a "before" and "after" resume sample.

RAY'S RULES OF RESUME STYLE

YOUR RESUME SHOULD BE TYPED, NOT TYPESET

You've selected high quality stock and had it engraved with your name, address, and telephone number. Your resume should be typed in black ink on that paper. The type should be no smaller than 12 pitch.

While some experts recommend professional typesetting, it has several disadvantages. It's expensive and it limits your ability to prepare, at a moment's notice, a resume targeted to a specific job. Typesetting gives the impression that you're doing a mass mailing, which, even if you are, shouldn't be apparent.

The only exception to the "don't typeset" rule is for those job seekers who have access to desktop publishing equipment with laser printers. This type of equipment combines the look of traditional typesetting with the flexibility and accuracy of word processing. You can create and store several versions of your resume and customize it at a moment's notice to suit a target job.

A cottage industry has sprung up in the past five years around this simple, relatively inexpensive technology. "Desktop Publishing" is now a heading in most Yellow Page directories. The quality of the services vary, so buyer beware. People from all backgrounds have quit their jobs and hung out "desktop publishing" shingles. Knowing how to work the equipment doesn't make them good typesetters. If you find a service staffed by former traditional typesetters, you've probably found someone with a good knowledge of typefaces, layout, and design.

If you go the desktop publishing route, select a traditional serif typeface (Times Roman, Century Schoolbook, Palatino) that matches or complements your stationery's imprint. The main text should be no smaller than 10 points on 12 points of "lead" (the space from the bottom of one line to the bottom of the next). Headings can be set in larger type.

If you don't use desktop publishing, your resume should be typed on a self-correcting, electronic typewriter with a carbon, not a cloth, ribbon. The typewriter should not have any broken elements or any other nuances that interfere with your resume's perfectly polished appearance.

Your best bet is a memory typewriter or word processor with a letter-quality printer. Once your resume or resumes are written and stored, they're available to you at the touch of a button. Typing each one in original is time-consuming and leaves too much room for error. If all this computer technology is foreign to you, it's a good time to make friends with it. You'll be acquiring a new skill to sell.

If you must rely on someone else to help you produce your resume quickly, make sure you get top-notch service. Don't trust another with its style and content; resume-writing services are out. Write and design it yourself and hire someone else to produce it according to Ray's Rules. If your spelling, punctuation, and proofreading are not perfect, hire a professional to do that, too. Typing services may or may not offer adequate help with this important final step.

BE GENEROUS WITH WHITE SPACE

Resume readers like to make notes, so give them at least a one-inch margin all around for that purpose. Don't crowd your text just to conform to the one-page rule. If you really have that much that must be included, go to a second page.

EMPHASIZE IMPORTANT POINTS JUDICIOUSLY

Boldface type can make important points stand out and grab the reader's attention. Just don't use this technique on every line or it loses its effect. Don't use words with all capital letters in the text portions of the resume. They're hard to read and detract from the overall look of the page.

Heavy blocks of text turn readers off. Wherever possible, use shorter paragraphs followed by bulleted lists of one or two line entries.

DON'T "X" OUT OR WRITE OVER UNDER ANY CIRCUMSTANCES

Recycled resumes are rotten resumes. If you don't display the energy and enthusiasm to produce a new resume, how can you make anyone believe you'll have it for the job?

Erasures, handwritten notes, white-out, correcting tape and any other clumsy corrections are also resume suicide. Truly, appearance counts.

DON'T INCLUDE A PHOTOGRAPH

This is only expected and accepted for fashion models, actors, and others whose looks are important to the position. For any other job, it raises eyebrows and lowers expectations.

That's it for Ray's Rules of Resume Style. Now let's cover content.

RAY'S RULES OF RESUME CONTENT

DON'T LIST AN OBJECTIVE

Confining yourself to one objective limits your opportunities. A better use for prime space at the top of the page is a Career Summary, which uses strong, eloquent language to summarize your career accomplishments and credentials.

LIST PROFESSIONAL EXPERIENCE IN INVERSE CHRONOLOGICAL ORDER

Under the heading "Professional Experience" or "Professional Achievements" list name of employer, city, state, dates of employment and title. Underneath each heading comes a position summary followed by bulleted list entries of significant accomplishments.

EMPHASIZE ACHIEVEMENTS AND RESULTS AS WELL AS EXPERIENCE AND RESPONSIBILITIES

"Experience and responsibilities" tell what you got paid to do and what you were supposed to do. Achievements and results, however, demonstrate what you have done.

BEGIN SENTENCES WITH STRONG, ACTION VERBS

Leave out "I." Everyone knows who you're talking about. Begin sentences with strong, action verbs that demonstrate high-level activity and achievement. Vary your language to keep it interesting. Don't ramble—sentences should be crisp, clear, and to the point.

QUANTIFY

Wherever possible, use numbers, dollars, and percents to convey at what level you operated and the significance of your achievements. "Directed large division of major manufacturer" doesn't say it as well as "Directed $100 million office furniture division of Fortune 500 manufacturing company."

INCLUDE EDUCATIONAL CREDENTIALS

Only the most important, such as college degrees, special certificates, and licenses. If you're a college graduate or have any postsecondary training, don't list your high school diploma.

INCLUDE AWARDS AND ASSOCIATIONS IF SPACE PERMITS

Prestigious awards and official positions held in professional associations can be included if space permits. If you have a laundry list of such accolades, however, don't try to squeeze them in. Under the heading "Awards and Associations" write "See addendum" and attach a separate list. The same rule applies if you've had many articles published in trade and professional journals.

DON'T INCLUDE REFERENCES

A simple statement "References available upon request" can be included, but if it isn't, it will probably be understood. The resume stage is too soon to unleash your arsenal of influential supporters.

TELL THE TRUTH

There's a world of difference between selecting language to present yourself in the best light, and lying. Never present inaccurate, false, or misleading information. It can and will be checked. If something you wrote doesn't check out, you'll be checked off.

DON'T GIVE A REASON FOR LEAVING PREVIOUS JOBS

There will be time enough to explain it all in the interview. It's not important on the resume, and it could get you disqualified.

DON'T STATE SALARY

Most advertisements ask for a salary history and most employers really don't expect you to give it until you start talking face to face. Leave off past as well as "desired" salary. It all depends on the job, after all.

SAMPLE RESUME: BEFORE AND AFTER

Jack Washington is an over-50 job seeker whose first attempt at resume writing was not bad. When he followed Ray's Rules, however, his second try produced a much more persuasive resume. Read the "before" version on pages 132–133: Can you discover the over-50 giveaways?

Before:

Jack M. Washington
939 Warren Street
Empire, Michigan 48084
(313) 551-0230

OBJECTIVE: Senior management position in which I can fully utilize the expertise gained in 30 years experience in machine manufacturing.

PROFESSIONAL EXPERIENCE:

General Machine Company, Ada, Michigan
1978-1990 MANAGER, MANUFACTURING OPERATIONS

Directed manufacturing for major manufacturer of machine shop equipment. Complete responsibility for production planning, material handling, metal forming, and assembly of a wide variety of machines. Directed a work force of 300.

—Consolidated assembly operations resulting in cost savings of $700,000 while achieving a 30% productivity improvement.

—Upgraded shop routings and order planning systems, resulting in $50,000 cost savings.

—Developed, with assistance of a community college, a labor skills assessment and training program which increased productivity, improved quality, and enhanced employee satisfaction.

—Planned and implemented cost reduction program to support business decline, resulting in cost savings of $8 million.

—Developed capacity and capability program for key customers, resulting in sales increase of $4 million.

—Front line control program to reduce labor costs. Savings $500,000 per year.

—Developed and negotiated with union a major job classification restructuring, which reduced classifications by 70% and saved $1 million per year.

—Implemented statistical process control (SPC) program for grinder operators, eliminating inspectors. Saved $200,000 per year.

JACK M. WASHINGTON   Page 2

—Developed and implemented an absentee control program
for hourly employees, saving $240,000 per year.

—Implemented energy conservation programs, reducing costs
by $450,000.

Blackstone Press Co., Detroit, Michigan
1975-78 SUPERVISOR, PRODUCTION AND INVENTORY CONTROL

Responsible for all production planning and inventory
control functions in plant manufacturing presses for the
kitchen appliance industry.

—Reduced standard parts backlog by 40%.

—Reduced inventory by $2 million.

—Installed new shop feedback system which cut costs by
$400,000 per year.

—Developed a timeshare planning system which enabled
staff to handle 100% increase in volume without adding
any personnel.

—Implemented over 20 production planning and shop routing
programs.

—Implemented a program to reduce customer return
inventory, saving $200,000 per year.

Walker Tool Company, Royal Oak, Michigan
1960-1975 GENERAL PRODUCTION FOREMAN

—Reduced scrap by 15% in a one-year period

—Developed and maintained reputation for being hands-on
foreman who could effectively train production workers
and get the work done through his people.

—Consistently exceeded quality assurance standards.

—Member of team which developed Walker's first inventory
control system.

MILITARY EXPERIENCE: U.S. Army Reserves, Corporal, 1960-64

EDUCATION:
B.S., Mechanical Engineering, Oakland University, 1963

After:

Jack M. Washington
939 Warren Street
Empire, Michigan 48084
(313) 551-0230

Career Summary:

Fortune 500 manufacturing executive with proven record of
success in production, production control, materials
management, and manufacturing engineering.

Strong planning, organization, communications, and human
resources management skills.

Outstanding track record of innovation in cost reduction
and productivity improvement programs.

Professional Achievements:

1978-1990
General Machine Company, Ada, Michigan
Manager, Manufacturing Operations

For $100 million division of Fortune 500 company. Managed
production planning, material handling, metal forming, and
assembly of custom stamping machines and presses.
Directed a management team of 12 and a work force of 300.

—Consolidated assembly operations resulting in cost
savings of $700,000 while achieving a 30% productivity
improvement.

—Upgraded shop routings and order planning systems,
resulting in $50,000 cost savings.

—Developed, with local community college, and instituted
a labor skills assessment and training program which
increased productivity, improved quality, and enhanced
employee satisfaction.

—Developed capacity and capability program for key
customers, resulting in sales increase of $4 million.

—Reduced labor costs $500,000 per year through front line
control program.

Jack M. Washington   Page 2

—Negotiated with union a major job classification
restructuring, which reduced classifications by 70% and
saved $1 million per year.

—Implemented statistical process control (SPC) program
for grinder operators, eliminating inspectors. Saved
$200,000 per year.

—Developed and implemented an attendance incentive
program for hourly employees, saving $240,000 per year.

—Devised energy conservation programs, reducing costs by
$450,000.

1975-78
Blackstone Press Co., Detroit, Michigan
Supervisor, Production and Inventory Control

For $50 million/year private company manufacturing presses
for the kitchen appliance industry.

—Reduced standard parts backlog by 40% and inventory by
$2 million.

—Installed new shop feedback system which cut costs by
$400,000 per year.

—Created a timeshare planning system which enabled staff
to handle 100% increase in volume without adding any
personnel.

—Devised, implemented and controlled more than 20
production planning and shop routing programs.

—Reduced customer return inventory, saving $200,000 per
year.

Other Experience:

Production engineer, production foreman, assembly worker

Education:

B.S., Mechanical Engineering, Oakland University

Now use your skills inventory, targeted career objective, and Ray's Rules to create *your* resume. When it's done, critique it using the checklist in Figure 9–1.

While your resume should be as persuasive, powerful, and perfectly prepared as possible, it's not the be-all and end-all. Face-to-face contact is more important. The over-50 applicant must meet with the decision maker to sell his or her skills, knowledge, and attributes. Resumes don't result in jobs. *Interviews* result in jobs. Chapter 10 discusses the many ways to use your resume, along with other techniques, to get those interviews.

Figure 9–1 Resume Checklist

RESUME CHECKLIST

☐ One page, two pages maximum
☐ Clean type
☐ Good balance
☐ Adequate white space
☐ One-inch margins
☐ Chronological format, most recent position first
☐ Strongly worded "Career Summary" at top
☐ Reads easily
☐ Crisp, clear sentences
☐ No long paragraphs
☐ Bulleted lists
☐ Important points highlighted in boldface (not too many)
☐ Dates, titles, and employers discernible at a glance
☐ No spelling, grammar or punctuation errors
☐ Uses action words
☐ Accentuates positive
☐ Stresses results over responsibilities
☐ Stresses accomplishments over experience
☐ Uses dollars, percentages, numbers
☐ Is a persuasive "preview of coming attraction"
☐ No salary data
☐ No "reasons for leaving"
☐ No dates or personal data that give clues to age
☐ No misstatements or exaggerations

10 | Getting Interviews

You are almost through with this book, and so far you've been occupied with what I call "inner-directed" activities. Part I considered the social and economic realities of finding a job after 50.

Part II guided you through all the important "pre-campaign" steps of inventorying your skills, targeting your objectives, and researching your targets.

Part III began with a chapter on writing your resume. So far, your job search has been a lonely one—thinking, researching, collecting data, writing. Now it's time to get out and meet people—many people.

In the course of prospecting for job leads, you need to make as many contacts as possible. From here on, when you're not out meeting people, networking, and interviewing, you should be talking on your personal "hotline" setting up appointments and interviews.

The more people you meet and talk to, the greater your chances of getting a job. Classified advertisements and employment services are the sources most people turn to first, but they open the door to only about 20 percent of the job universe. Eighty percent of all jobs are filled without first being advertised to the general public or listed with an agency. They're won by people who knew people who knew people—and who finessed their way to an interview *before* anyone else got there.

GET WORKING NETWORKING!

Networking, which is talking to the people you know and to the people *they* know, takes on a unique slant after the age of 50. Most 50-plus job seekers have a wider range of personal and professional contacts than their younger colleagues have. Not only do they know *more* people, but the people they know are *more influential*.

Networking, therefore, should be a piece of cake, right? Two or three phone calls and you're on your way!

Unfortunately, it's not always that easy. Many people over 50 are uncomfortable with typical networking activities. Some are accustomed to being the leaders to whom others come for guidance and influence. Others are just uncomfortable with discussing their situation. They're embarrassed to admit being out of a job.

While discomfort and embarrassment are normal reactions, they are also counterproductive. And they must be overcome. To find a job successfully after 50, you must unabashedly and self-confidently activate your network, investigate available opportunities, and use whatever influence you can uncover to get your foot in the door.

You're not asking for a handout. When you've landed your new position, you'll be ready to assist others, won't you? Reciprocity. Mutual exchanges. Cooperation among colleagues. That's how networking works, and it makes the world of work go 'round. So, get ready to grab your brass ring on the career carousel.

MAKE A LIST

Begin by making one last list—everyone you know who might be a link to a job *opportunity*. The people on your list don't have to be in a position to hire you, or even aware of openings in their companies. They just have to supply an inside track, through themselves or others, to potential opportunity. Thousands of jobs are *created* every month when the right person comes along. Networking leads you to people with needs. After that, it's up to you to show them how you can meet those needs.

The best networking contacts are those who are concerned about you, willing to assist you, and know others to refer you to. Consider any or all of the following:

‖ Current and former colleagues

‖ Current and former supervisors

‖ Current and former subordinates

‖ Current and former clients/customers

‖ Current and former suppliers

‖ Friends who are employed

‖ Friends who know others who are employed

‖ Neighbors

‖ Clergy and members of your church or synagogue

‖ Fellow members of civic organizations

‖ Fellow members of trade and professional associations

‖ Fellow members of clubs

‖ Spouse's colleagues and friends

‖ Relatives

‖ "Personal" professional contacts (e.g., your doctor, lawyer, insurance agent, stockbroker, accountant)

‖ Editors of trade journals (they know *everyone*)

‖ Officers of trade and industry associations

‖ The career placement office at your alma mater.

You may have even more sources of contacts. Go through your address book, Christmas card list, old files and memos, anything that will jog your memory about places you've been and people you've met along the way.

Recall business trips, seminars, trade shows. Dig out the business cards of acquaintances you always meant to keep in touch with. Maybe you're the kind of person who *did* keep in touch. If so, you've been networking all along, and this step in the job search will come easily to you.

Once you've made your list, reorganize it from "Most likely to help" to "Least likely to help." Some people are, by their nature, born networkers. They take a genuine interest in others, remembering names and details. If you encounter anyone like this, chances are they'll reel off a list of names that you should know. People who know you well may want to help but simply don't have the contacts to help you. Others may not know you well but have a wealth of information and contacts.

When your list is in priority order, roll up your sleeves and start calling. *Keep* calling until you connect. Get as much face-to-face contact as you can. Set up lunch and breakfast meetings or arrange to "drop in for a few minutes."

As you activate your network, you're sure to come across information that may not be immediately useful to you but which could help someone else now or in the future. The best networkers treat all information, every name and phone number, as precious resources that become valuable when traded. Fine tune your information sensors. Know a good lead—for yourself or another—when you see it, and make the most of it.

PREPARE A SCRIPT

It may be difficult to pick up that phone and start talking; but, with practice, prospecting by telephone gets easier. To overcome anxiety, it helps to have a script in front of you for guidance. Eventually, you'll be persuasive, self-confident, and captivating in your calls. (And you'll be getting great practice for your face-to-face meetings.)

If you've targeted a specific company but do not know anyone there or anyone who can refer you, "cold" call the company and ask for the department head or heads related to your work, such as Plant Manager, Marketing Manager, and so forth.

Your call should be brief, and its purpose should be to set up an interview with a decision maker. Without reeling off your whole resume, briefly describe yourself. Mention two or three selling points:

> Mr. Smith, this is John Jones calling. I worked for ABC company until they closed their plant last month. My most recent title was Quality Control Supervisor, but I have extensive general management experience and skills that would transfer with ease to another operation.

Even if there are no current openings, you would like the "opportunity to discuss how my experience and knowledge may benefit XYZ now or in the future."

Be careful not to position yourself as "old." Don't say you have "30 years experience" or you "worked there all my life." Show that you've kept your skills up to date, and demonstrate your knowledge of the prospect's company.

> I recently completed a course in production robotics at the University, where, by the way, I earned an MBA in general management two years ago. I'm interested to learn how XYZ's new robotics project is going.

Take the initiative and ask for an interview. Don't ask a question that could be answered "yes" or "no."

> When would be a convenient time for me to come in to see you?

At the very least, try to get the name or names of other potential leads. Follow up every call either with a note confirming the date and time of your meeting or a resume with a cover letter that expresses appreciation for the time and information and mentions one or two more key points about you. Be (politely) persistent. The meek may inherit the earth, but they don't often succeed at getting appointments by telephone.

MORE TELEPHONE TIPS

▪▪ Call between 7:30 and 9:00 A.M., when the manager is fresh and before he or she gets embroiled in the day's activities. If you call before 8:00 A.M., there may be no "secretary screen" set up yet to stall you. Several of the people I have counseled have also been successful in getting through to the decision maker on Friday afternoon.

▪▪ If the switchboard is not open early, try to call the day before to ascertain the decision maker's name and direct dial number.

▮ Place a mirror in front of your telephone and check to make sure you're smiling while talking. A positive expression will come through in your tone of voice.

▮ Before talking with a woman, first learn if she is addressed as "Ms.," "Mrs.," or "Miss." *Do not* assume a female voice is a clerical person.

▮ Enlist the secretary as your ally. Don't ever annoy or bully him or her. The secretary is not an obstacle but a vehicle through which you can reach the person in authority. Learn the secretary's name, write it down, and use it on subsequent calls.

▮ Put authority and confidence in your voice, and you'll be put through.

Let's listen in on a call. Roberta Able is calling Intrepid Industries.

Receptionist:	"Good Morning, Intrepid Industries."
Roberta:	Good morning. Can you tell me who is in charge of pharmaceutical sales?
Receptionist:	Pharmaceutical sales? That would be Pat Baker.
Roberta:	Is that Mr. Baker?
Receptionist:	Oh, no! Pat is short for Patricia!
Roberta:	May I speak with Ms. Baker, please?
Receptionist:	I'll connect you.
Secretary:	Pharmaceutical Sales.
Roberta:	Good morning. This is Roberta Able. May I speak to Ms. Baker?
Secretary:	Thank you. (Silence, while secretary checks. Then,) May I tell Ms. Baker what this regards?
Roberta:	Please tell Ms. Baker that I have been in pharmaceutical sales with A.L. Laboratories and I'd like to speak with her about career opportunities with Intrepid. (Or, better

yet: "Please tell Ms. Baker that Homer Garth recommended I call her.")

Secretary:	One moment.
Pat:	Pat Baker.
Roberta:	I'm Roberta Able, Ms. Baker. As you know, A.L. Laboratories has downsized its sales force. I'm one of the people caught in the crunch. I have in-depth experience in pharmaceutical sales.
Pat:	I'm sorry. I'm not looking for additional salespeople at this time.
Roberta:	I can understand that. I would like to meet with you to tell you about a unique method I developed for increasing sales to independent pharmacies. It helped make me the top producer in my division at A.L. for five years. My degree in pharmacy helped me develop a special rapport with drugstore owners.
Pat:	It sounds like A.L. lost a top employee. Why don't you send me your resume?
Roberta:	I'd like to drop it off and meet you. How would Tuesday morning be?
Pat:	Tuesday at ten would be fine.
Roberta:	Thank you. I'm looking forward to seeing you.

CLASSIFIED ADS

Once a help-wanted ad appears, you've got lots of competition. Answering ads is not the most effective way to prospect, but it's a method you cannot ignore. Set aside some time daily for this task. A list of guidelines for maximizing your success with classified ads follows.

RAY'S RULES FOR RESPONDING TO CLASSIFIEDS

1. Read all the ads, from A to Z. A position may be listed under almost any title. There are no rules for alphabetizing them. For example, a secretarial position may be under A for administrative assistant, E for executive secretary, or S for secretary.

2. Don't just answer the ads for jobs that match your qualifications. (No one's qualifications *exactly* match the stated requirements, ever.) Human resources and the department doing the hiring might not even be in close communication, and the requirements listed in the ad may not exactly match the manager's wish list.

   The mere existence of an ad, furthermore, could be an indication of an expansion, reorganization, or other company activity that will result in more opportunities. Sometimes, the position has already been filled—from within, leaving an opening elsewhere.

3. Look at the *company* that is advertising, not just the position advertised. Do some quick research to scope out their operations. Would they have *any* positions for you?

4. Call the company to get the name of the decision maker in charge of hiring for the position.

5. If you know someone who knows the decision maker, ask them for a written or telephoned introduction for you.

6. Call and speak to the decision maker without delay. Inquire in general about operations. With some active listening, you might learn of other openings or, at least, opportunities.

7. Try to get an interview on the spot. If they say, "Send a resume and we'll get back to you," forward one immediately with a cover letter that references your phone call. Send it to the person you spoke to. Make sure you get the correct spelling of the individual's first and last name, title, department, and address.

8. Call to follow up, within three days of your resume's arrival, to schedule an interview. If they put you off,

you might tell them you are being pressured to respond to another offer, but that their position sounded like such a good match that you wanted to explore it before making a decision. In the meantime, keep networking for contacts who might be able to wield influence for you.

9. Finally, if you are completely unable to turn up any information on the position, send your resume anyway to the address in the ad. (If you can, get the name of a personnel officer from the department's receptionist. Anything that personalizes your envelope will give you an edge over all those envelopes addressed to "Employment Manager.")

10. Study the advertisement carefully for clues, and use any information you may have gleaned from telephone investigations, to write a resume tailored to the position.

11. Call to follow up all resumes sent. Don't be easily dissuaded. Your diligence will break down barriers, and it will set you apart from the majority of job seekers who easily take "no" for an answer.

12. Always be pleasant and polite. Snarl at a secretary just once, and I guarantee you it will be the last time you get through.

13. If you're currently employed, avoid answering "blind box" advertisements that don't reveal the company name. Some are phony. Others are real, but you never know when you might be applying to your own employer.

CLASSIFIED AD-VANTAGE

Networking can work wonders when answering an ad. Instead of adding your resume to the pile of unopened unknowns in the personnel office, network your way over the heap and straight to the decision maker. Here's how one tenacious over-50 job seeker did just that:

Mark, 52, was the director of operations for a manufacturing company when a buyout ushered in a new management team. Mark was ushered out with six months' severance and a handshake.

He went right to work and was a dedicated, thorough job searcher. He did everything right, but, six months later, his severance was running out and, although he'd had many interviews, no worthwhile offers had been extended. His wife's career and his children's activities made relocation a last resort, but in the Northeast in the late 1980s, manufacturing management positions of the caliber Mark sought were few and far between.

Mark was beginning to lose hope when an ad appeared in the Employment Classifieds one Sunday for industrial engineering manager of a production facility in a nearby town. It was a lower level position than the one Mark had left, but he was definitely interested. Mark used his networking contacts to bypass personnel and learn the name of the individual in charge of hiring. First thing Monday morning he spoke to the plant manager, and by Monday afternoon had dropped off a resume.

But, when Mark called to follow up, he was stonewalled. The manager claimed they were looking for someone with an engineering degree. Although Mark had been an engineering manager for years before being promoted to operations director, his degree was in business administration. Mark's attempts to convince the company to interview him were unsuccessful.

Mark was discouraged, but not defeated. Determined to have his "day in court," he went deeper into his network. Instead of just getting names this time, he uncovered contacts who actually worked at the company or knew people who worked there. A friend from his college days worked in accounting. Mark's wife's former boss lived in the same neighborhood as a vice-president of the company. Together, Mark and his wife prevailed on their contacts to "put in a good word" for Mark, and he got an interview.

When Mark met face to face with the plant manager and others at the company, they developed instant rapport. On a tour of the plant, Mark demonstrated his superior industrial engineering management knowledge. The company was so eager to have someone with such high level background in their small but rapidly growing plant that salary range standards went out the window. Mark's job offer came with a salary and bonus program that almost matched what he had been making, with plenty of room for growth.

If Mark had merely sent his resume to the company's personnel office, he would have been disqualified automatically based on his educational credentials alone. When the personnel department is told to look for specific credentials, only the resumes showing those credentials get passed along for review. The rest get a "no interest" letter.

Over-50 job seekers have to use their cunning, and their connections, to meet face to face with the decision makers. Your superior experience, skills, and presence will shine through in an interview in a way they never do on paper and rarely can on the telephone.

CREATIVE COVER LETTERS

With all the leads you'll turn up through your prospecting activities, you're going to be forwarding a lot of resumes with cover letters. Don't settle for the ordinary "enclosed please find" cover letter style. Make your cover letters an adventure in reading. Use them to:

1. *Remind* the reader of your recent telephone conversation.
2. *Reveal* the name of your mutual associate or friend.
3. *Refer* to specific notes of interest in your attached resume.

Cover letters should be:

1. No more than one page in length;
2. Typed, on engraved stationery, in businesslike block format (see samples);
3. Completely free of spelling errors or erasures;
4. Addressed to a decision maker whose correct name and title are *used and spelled correctly*;

5. Well written and interesting to read;

6. Followed up with a telephone call.

The body of a good cover letter consists of three parts:

▮ *The opening*—a brief one or two-sentence paragraph that states your purpose. It should grab and hold the reader's attention.

▮ *The persuasive paragraph*—a longer paragraph that sells the reader on your unique skills, abilities, and the contributions you can make to his or her department.

▮ *The closing*—a brief, final statement which asks for action or tells them what action you will take: "Kindly call at your earliest convenience," or, better, "I will call within a week to arrange a time to meet and discuss this."

Some examples follow on pages 149–150.

STEVEN J. MARWICK
431 W. 102nd Avenue
Los Angeles, CA 92005
(213) 786-9089

August 18, 1990

Anne L. Godwin, Employment Manager
Cartech, Inc.
701 W. 27th St.
Modesto, CA 95351

Dear Ms. Godwin:

Recently I learned through Bill Kennedy, your General
Accounting Supervisor, of your company's plans to expand the
production facility at Modesto. I am interested in the industrial
relations opening he mentioned.

As my enclosed resume will reveal, I have in-depth experience
in human resources management. I have worked in a plant
with over 1,000 employees where I headed up an employee
relations program that reduced grievances by 50 percent.
My successful work in employment, training, benefits, and
compensation have rounded out my personnel management
skills.

I'd welcome the opportunity to discuss how my skills and
experience can make a valuable contribution to Cartech's
operation in Modesto. I'll call within the week to set up a
definite appointment. I appreciate your consideration and
look forward to meeting you.

Sincerely,

Steven J. Marwick

ELLEN SMITH
104 S. Oak Street
Columbus, Ohio 32145
(901) 645-7856

May 18, 1990

Patrick J. Henry, President
Premiere Graphics Corporation
25 Industrial Parkway
Kansas City, MO 64537

Dear Mr. Henry:

Our mutual friend, Judge Samuel Siebert, told me you were looking for an executive secretary. I am planning to relocate to Kansas City next month and am interested in the position.

Along with the resume I've enclosed is a letter of recommendation from Judge Siebert. You'll note in my resume that I have 15 years' experience in legal and executive secretarial positions, type 95 words per minute, and take shorthand. Supervising and organizing an office are my strengths. I enjoy a busy work environment and love the challenge of learning new things. My employers have always been able to delegate important responsibilities to me with complete assurance they would be handled competently and efficiently.

I will be visiting Kansas City on a house-hunting trip next week, and I would like to meet with you then if it is convenient. I'll call you at the end of the week to set a definite time. In the meantime, thank you for your consideration. I look forward to meeting you.

Sincerely,

Ellen Smith

OUTSIDE SERVICES

If you follow all the suggestions and guidelines in this chapter for uncovering job leads and getting interviews, you will probably be so busy you won't need to seek assistance from outside employment services. If that is not the case, however, or if you just want to make sure you've got all your bases covered, I offer you the following brief primer on the industry. As in all things, an informed buyer makes better choices.

CAREER COUNSELING SERVICES

Ads for these services can be found in the classified section of Sunday newspapers alongside the help wanted ads. Not employment agencies, these "counselors" charge you $2,500 to $5,000 and up (Visa and MasterCard accepted!) for advice. Most give each client a standard battery of aptitude and interest tests, help you put together your resume, and run through practice interviews with you (with or without video playback). They'll give you books like this one to read.

Save your money. Nearly everything they'll tell you about the job search is already in this book. And it's time consuming. Don't get bogged down—you need to be on the phone and out in the world making contacts.

OUTPLACEMENT

Outplacement services are growing, in scope and in stature. Although individuals can contract with an outplacement firm and pay the fee, most outplacement firms are paid by employers who are terminating employees. Among other services provided, outplacement firms:

■ Counsel an about-to-be-terminated employee.

■ Counsel on issues such as job targeting, relocation, self-employment, job prospecting, and salary negotiations.

■ Assess the employee's job strengths and weaknesses.

- ▮ Identify the employee's transferable knowledge and skills.
- ▮ Develop a job seeking strategy.
- ▮ Help prepare letters and resumes.
- ▮ Target-mail letters and resumes.
- ▮ Improve the candidate's interview skills through videotaped practice interviews.
- ▮ Provide office space, research materials, and secretarial services.
- ▮ In some cases, provide actual job leads.

If your employer offers outplacement services at no charge to you, by all means accept. Your outplacement team will provide expert advice while giving you a businesslike base of operations. (It is much better to conduct your job search from the outplacement firm's offices than from your "old" office. If your employer offers use of an office and telephone, turn it down. It is demoralizing to see your "former" coworkers still at their jobs when you're no longer "part of the team.")

Realize, however, that the outplacement firm is not getting you a job. You are getting you a job. Your counselors are working with you to organize and manage your job search so that you achieve success in the shortest possible time.

Beware of firms that advertise themselves as outplacement firms when they really are either contingency employment agencies or career counselors. It's outright misrepresentation, and it's done to confuse unsuspecting job seekers by aligning with the reputable, employer-paid segment of the industry. Outplacement firms do not advertise to the general public or solicit resumes from people seeking employment.

EMPLOYMENT AGENCIES

There are two types of employment agencies, differentiated most clearly by how they are paid. The two types are "retained" recruiters and "contingency" recruiters.

∎ *Retained recruiters* are paid by the employer even if the candidates they refer are not hired.

∎ *Contingency recruiters* are paid by the employer only if the candidate they refer is actually hired.

A few agencies are paid by the job seeker. Be certain that you are referred only to "employer-paid" jobs.

Retained recruiters are known by such names as executive recruiter, management recruiter, technical recruiter, executive search consultant, and management search consultant. Employers retain them to conduct a search for high level executives—usually in the $75,000+ brackets.

Retained recruiters are in the intelligence business. They're sleuths who prefer to discover you before you discover them. You can send them your resume and reference letters with a cover letter, then follow up with a phone call to see who is most interested. If the recruiter doesn't contact you or return your phone calls, give up. He or she isn't interested, and you're wasting your time.

Check your network to see if anyone can recommend an executive recruiter. Recruiters are expert networkers. They'll be more likely to listen to you if you drop the name of someone they know.

Contingency recruiters are more likely to call themselves "employment agencies," but to confuse the issue they may also call themselves search consultants, technical recruiters, and so on.

Employment agencies are more effective for job seekers in the early stages of their career (entry level to middle management), in the $30,000 and under salary range, or who have "hard" skills such as secretarial, engineering, and accounting. If you fit their profile and are interested in learning what they have to offer, you'll find them in the *Yellow Pages* under "Employment Agencies," "Personnel Agencies," "Personnel Services," "Employment-Permanent," or similar headings.

TEMPORARY EMPLOYMENT SERVICES

I talked about the new wave in temporary employment in Chapter 6, "Targeting the Right Job." "Temp" and "Exec Temp" services are paid by companies to provide "fill-in" help. They then pay the "temp" an hourly rate that is somewhat lower. The service makes all employee payroll deductions and pays employment taxes like any other employer. Some even offer health insurance benefits, paid vacations, and bonuses.

As recently as 1980, most temporary services specialized in general labor or clerical jobs. But as employers recognize the high cost of carrying full-time employees on the payroll, they're using temps for interim assistance in everything from accounting to engineering to executive leadership.

As outlined in Chapter 6, "temping" can be an ideal avenue for the over-50 job seeker, either "temporarily," or permanently. Check temp agencies as thoroughly as you would other employment agencies.

STATE JOB SERVICES

Free to job seekers, these services list job openings. Some have counselors who will work with you. Those receiving unemployment compensation usually must register with their state job service. It doesn't hurt to check the listings, but don't over-rely on them.

FORTY-PLUS CLUBS

This is a nationwide network of nonprofit, corporate executive, job-support organizations staffed by volunteers. They are

> dedicated to helping unemployed managers, executives and professionals, 40 years of age or older, to conduct an effective job search campaign. Forty Plus has strict entrance requirements since it also functions as a quasi-executive recruitment organization, and wants to assure potential "clients" that the members it recommends for positions have a high standard of excellence.[1]

Current Forty-Plus Club Locations

California
7440 Lockheed Street
Oakland, CA 94603
(415) 430-2400

3450 Wilshire Boulevard
Los Angeles, CA 90010
(213) 388-2301

23151 Verduga Drive
Laguana Hills, CA 92353
(714) 581-7990

Colorado
639 East 18th Avenue
Denver, CO 80203
(303) 830-3040

3840 South Mason Street
Fort Collins, CO 80525
(303) 223-2470 Ext. 261

2555 Airport Road
Colorado Springs, CO 80910
(719) 473-6220 Ext. 271

Hawaii
126 Queen Street
Suite #227
Honolulu, HI 96813
(808) 531-0896

Illinois
53 W. Jackson Boulevard
Chicago, IL 60604
(312) 922-0285

New York
701 Seneca Street
Buffalo, NY 14210
(716) 856-0491

15 Park Row
New York, NY 10038
(212) 233-6086

Ohio
1700 Arlingate Drive
Columbus, OH 43228
(614) 275-0040

Pennsylvania
1218 Chestnut Street
Philadelphia, PA 19107
(215) 923-2074

Texas
13601 Preston Road
Dallas, TX 75240
(214) 991-9917

3935 Westheimer
Suite #205
Houston, TX 77027

Utah
1234 Main Street
Salt Lake City, UT 84147
533-2191

Washington, DC
1718 P Street, N.W.
Washington, DC 20036
(202) 387-1582

OPERATION ABLE

This is a government-funded agency that specializes in placing older people, usually the economically disadvantaged and usually in lower paying service jobs. To locate a local office, contact your state office for the aging.

INFORMATION SERVICES

Finally, there are a few fly-by-night operators who merely list job openings. They charge job seekers from $75 to $90 for a copy of their "exclusive" list—which is usually no more than a compilation of classified ads. You probably know more about what's available than the list will tell you. These firms spring up in areas where unemployment is high or large layoffs have recently occurred. They're predators and quick-buck makers. Don't bother.

KEEP A LOG

Organization and follow-up are essential for effective prospecting. Before you launch your high-powered prospecting program, put together a system to help you keep track of your leads.

You can do this with simple 3 × 5 file cards (See Figure 10–1) on which you record name, phone number, and other important details. Organize them in a file box in any way that is pertinent to you—by company name, job classification, and so on. (Or keep all information on pages in a three-ring binder.) Take your prospecting notes with you when you lunch with a colleague, meet with someone for "information-gathering," or go on an official interview. They may come in handy.

Figure 10–1   Contact Log

Name: _____ Title: _____

Telephone Nos.: (H) _____ (O) _____

Company: _____

Address: _____

Referred by: _____ Contacted on: _____

Appointment scheduled for _____

Results: _____

_____

Other leads: _____

_____

☐ Resume sent on _____   ☐ Thank you sent on _____

This chapter provided a summary of on the many ways you can uncover job leads and get interviews. The most important thing to remember is that it's a numbers game. The more leads you track down, the more interviews you're likely to get. The more interviews you go on, the greater your chances of finding an opportunity that fits and wears well for you. In Chapter 11, I give you tips for increasing your interviewing success.

NOTE

1. Birsner, E. Patricia, *The 40+ Job Hunting Guide,* New York: Simon & Schuster, 1987, p. 241.

11 | Interviewing Successfully

When you study the books that professional employment people read on how to interview, one principle becomes obvious. The person interviewed does not receive an objective evaluation based on skills, experience, and knowledge. If that were so, there would be no need for an interview!

The interviewer judges you on *how you act* more than on *what you say.* You're giving a performance to a critical audience, and they're just waiting for you to miss a cue. In *The Complete Q&A Job Interview Book,* Jeffrey Allen says:

> A job interview is a screen test. An act. Getting hired depends almost completely on the "actor factor." If you know your lines, perfect your delivery, and dress for the part, you'll get hired. If you don't, you won't. No retakes. No bit parts.[1]

Interviews are *never* fair. If you expect them to be, then you'll be monumentally disappointed. In a subjective, selective, one-hour session, someone will determine whether or not to enter into a long-term relationship with you. (Or, in the case of a first, or screening, interview, they'll decide whether you will "pass" to the next test.)

There is no way anyone can know *you* in an hour, so you will be judged on what you *project.* Physical appearance, mannerisms, gestures, poise, confidence level, personality traits, how well-prepared you are, how smoothly you field difficult questions—all these factors contribute to your "score." Interviewers want to know:

∎ Was your resume a professional and accurate representation of pertinent facts?

∎ Was your appearance neat and businesslike? (Did you "dress the part?")

∎ Did you grasp readily the essence of the questions asked?

∎ Did you answer tough questions without hesitating, and with sincerity and conviction?

∎ Did you have all necessary information available (names, dates, references, etc.) when requested?

∎ Were you prepared, confident, and self assured?

∎ Did you arrive on time?

∎ Were you convincing?

∎ Will you fit in?

An interview really begins long before you walk through the door. All of the steps in your campaign—from inventorying your skills through targeting your objective, researching your targets, writing a resume, identifying contacts, and setting up interviews—help you prepare for this final hour.

GET "PSYCHED"

Your hard work will go to waste, however, if you don't develop the right *mindset* for interviewing. Over-50 job seekers are particularly vulnerable in interviews, because they often adopt one of two dangerous attitudes: Here's the first:

I've been at this career so long, I talk about it in my sleep. I know my stuff. I don't need to prepare!

If that's your approach, prepare to enjoy an early and long retirement. Then there's:

> I held my last job for 15 years. I don't even remember my last interview. I *can't* interview.

Then you *can't* get a job. *Anyone* can interview successfully. It requires understanding the process, and *being prepared*. In other words: Know the rules, and play to win.

BEFORE THE INTERVIEW—BE PREPARED!

Make the Boy Scout motto *your* motto. The more prepared you are, the more in control you'll be. You'll never be completely in control, but you'll be able to sway the odds in your favor.

SCHEDULE FOR SUCCESS

When you interview can make or break you. You want to be fresh and at your best, and you'd prefer that the interviewer not be distracted. It's easier to control the former than the latter, so make sure you don't overload your interview schedule.

Try to schedule two interviews per day, the first at 9:00 or 10:00, and the second at 2:00. This keeps you on your toes without wearing you out. You get enough of a break between appointments to refuel, recharge, reread your research on the next company, and get mentally ready. Try to do this at least three—if possible, four—days per week, with one day left for follow-up telephone calls, letters, and scheduling more interviews.

This timetable also helps you avoid eating a meal with the interviewer. Eventually, as you progress to second and subsequent interviews, you'll probably have to sit down to lunch or dinner with a decision maker. But, in the early stages of the process, too much can go wrong. Remember, the interviewer is making snap judgments. Use the wrong fork or spill your drink and your audition may end with the words, "Don't call us, we'll call you."

RESEARCH THE COMPANY

When you have an interview arranged, review your research notes to learn everything you can about the company. Call your contacts to see what they know. Make anonymous telephone calls to the company to gather more data.

You're looking for more than product lines and market share. You want to know about the company culture, its values, and the kinds of employees it hires. In your first interview, the employment official won't be testing your job skills as much as deciding whether or not you "fit in with the family." In *every* interview, you'll get high marks for doing your homework and knowing what the company is about.

DRESS THE PART

I talked about image in Chapter 4. Reread that section. If you are "showing your age," don't compound the problem with careless or outdated grooming and apparel.

"Looking old," being seriously overweight, appearing out of shape, (breathless after climbing stairs), all will be silently noted. They may go through the motions of an interview, but they've already decided against you. A vital, energetic, positive, and well-groomed approach is *essential* after 50. You can complain about the unfairness all you want. It won't change it. And you won't get a chance to change things unless you get the job, so play by the rules.

▪ Are your eyes clear? Get a good night's sleep and use eye drops.

▪ Is your weight proportionate to height? If, despite diet and exercise, the pounds are permanent, compensate with especially flattering, well-tailored clothing. You may have to pay more, but it's a career investment.

▪ Are you "covering the gray?" Just make sure it looks natural. Balding isn't a sin, either, unless you contrive to cover it with awkward hairstyles and obvious hairpieces. That's just inviting negative attention.

▌ Women, don't overdo with makeup. Like your hairstyle, it should be conservative. Wear little or no jewelry. Nothing should detract from *you*. Your interviewer should be looking at your face and listening to your voice, not staring at your diamond rings.

If you're uncomfortable about your appearance, it'll show. Prepare for your interview long before by beginning an exercise and healthy diet regime. You don't have to become a body builder. In fact, you shouldn't overdo. Moderate amounts of fresh air, exercise, and the proper food will make you appear more youthful, energetic, and self-assured.

BE ON TIME

Always leave in plenty of time to allow for traffic and other contingencies. If there's even a remote possibility of getting lost, make a "dry run" before your interview to make sure you know how to get there.

Don't, however, show up at the receptionist's desk more than five minutes before your interview. Arrive early, but use the extra time to check out your surroundings, make sure you know where you're going, and step into the restroom to freshen up before approaching the receptionist and announcing yourself.

If you're running late, call to let them know. If it will be more than 15 minutes, try to reschedule. Don't risk an irate interviewer.

APPLY IN ADVANCE

Once you're prepared for your interview, you don't want any distractions to deter you. And paperwork shouldn't interfere with first impressions. Filling out an application when the interviewer arrives can create an uncomfortably awkward introduction. If possible, get a copy of the application in advance, have it neatly typed and signed and hand it to the receptionist or interviewer when you arrive.

REMOVE SUNGLASSES AND OVERCOATS

You must be at your business best when you greet the interviewer, so remove sunglasses before you walk into the building and hang your coat up (or leave it on a chair in the reception area) before being escorted into the interviewer's office. Fumbling with coats also interferes with that vital first impression.

PACK YOUR BRIEFCASE

There are certain things to bring to each interview to produce if requested. (See Figure 11-1.) Carry them in a handsome, high-quality leather briefcase. (Like your shoes—polished and presentable.)

Figure 11-1   Interview Checklist

☐ Resumes—six copies on quality bond stationery (not photocopies)

☐ Professional references—names, addresses, and contact telephone numbers of three or four former supervisors and other professional references

☐ Letters of reference—These are optional. The best letters of reference are those written directly to a decision maker by an influential reference after you've been interviewed. But, if you have some recent "To whom it may concern" letters written by influential references, bring copies

☐ Samples of your work

☐ College transcripts

☐ Social security number, special licenses

☐ Names of all former supervisors; names and addresses of all former employers, with exact dates of employment

☐ Military discharge papers, letters of commendation

☐ Legal pad and two good ball-point pens

THE FIRST FIVE MINUTES

Most interview experts will tell you the first five minutes decide the interview. How you greet, shake hands, and get settled into your chair will set the stage for the rest of the interview. If you appear uncomfortable or awkward, you'll make a negative impression that will last for the rest of the interview and nothing you say will help.

Practice your initial greeting. Do it in front of a mirror, or get a friend to role play with you and another to videotape the session until it's perfect. If you rehearse, you'll be less nervous and therefore less likely to say something awkward or embarrassing. Tension makes fools of us all. Preparation eases tension.

GREET THE INTERVIEWER WITH A GENUINE SMILE AND A FIRM HANDSHAKE

Both men and women should shake hands when introduced, and both must have a firm handshake—not a bone crusher, but not a limp rag, either. A firm handshake is an American business standard. There are people who won't hire others who don't shake hands properly (and you thought *age* discrimination was unfair)!

Make sure you're smiling with both your mouth and your eyes. Don't strive to say something clever. Save your best stuff for when you're settled in and the interviewer is listening. Now the interviewer is concentrating on meeting you and learning your name.

Look the interviewer in the eye, extend your hand, grasp the interviewer's hand, shake firmly, and say,

Good Morning, Ms. Adams, I'm Timothy Eldridge. I'm pleased to meet you.

Speak slowly and clearly, especially if your name is difficult to pronounce.

BE FRIENDLY AND PLEASANT

This is a personality contest. Nice people win it. These seem like such basic tenets. You've been conducting yourself professionally for years, I know. But even seasoned actors never go before an audience without first making sure they can walk on stage and hit their mark. Leave nothing to chance. Rehearse.

RAY'S RULES FOR INTERVIEW SUCCESS

You're in. You've seated yourself, the interviewer has closed the door and walked behind the desk to sit down. Shortly I'll discuss the content of the interview, what questions to expect, how to answer the tough ones about age, and so on. But first there's a list of rules and techniques you must know. Here it is.

DO NOT CHEW GUM OR SMOKE

Even if you are invited to smoke, even if the interviewer smokes, don't. This isn't a social call. It's a presentation. Don't use valuable time lighting up, blowing smoke, and flicking ashes off.

Smoking is declining in popularity in the work place. An over-50 applicant who smokes is inviting the interviewer to make some assumptions about his or her health and potential longevity. (Don't have smoking materials or a lighter in evidence on your person or in your briefcase, either.) The smell of stale smoke on clothing and breath is a definite interview turn-off. If you smoke, use breath mints or spray before you enter the building. Smoking in your car or other confined area before the interview will make smoke cling to your clothing and hair.

ESTABLISH RAPPORT WITH THE INTERVIEWER

Once inside the interviewer's office, display your charm and personality. Look around. Note something on which you can comment. (Avoid personal items such as photographs.) Perhaps there's an award, or an architectural rendering of a proposed new facility, hanging on the wall. Make a brief comment, but avoid a long conversation. You're there to talk about *you*.

SPEAK CLEARLY, DIRECTLY, AND POLITELY

Prepare for each interview so that you don't get caught off guard. If the interviewer does throw a curve ball, think your answer through before speaking. Present your response directly and convincingly, using examples and numbers where appropriate. Keep your answers brief, but complete. Most interviewers can detect a "snow job." Be honest. If you don't know an answer, say so.

MAKE EYE CONTACT

Always look the interviewer in the eye. Interviewers are trained to pick up nonverbal cues. Failure to make eye contact could be construed as dishonesty.

EXPRESS ENTHUSIASM

Sit up, lean forward. Use positive body language. Don't slouch. What does your posture say about you? Confident? Interested? Bored? Tired? Nervous? Display an interest in the interviewer as well as the job.

BE POSITIVE

Put all experiences in a positive light. Don't criticize former employers or colleagues. Referring to "personality clashes" is interview poison. "Policy differences" is a much better term to use. Don't let bitterness or anger creep into your interview. A negative comment will reflect poorly on you, not your target.

POINT OUT PERTINENT EXPERIENCES

Sometimes you have to help the interviewer along. They don't always ask the right questions. While you shouldn't "tell a story" to illustrate *every* answer, take the opportunity when appropriate to further establish your qualifications.

BE AWARE OF WHICH INTERVIEW YOU'RE IN

There are first (screening) and there are second interviews. In some companies, there are third, fourth, and fifth interviews. Each step in the process is designed to narrow the field.

If you're being interviewed by the manager who will hire you, then your interview content will focus on the details of the job and your skills. If you're in an initial interview with the personnel manager, however, you're there to show what type of person you are. Don't bore the personnel person with a long-winded explanation of the last computer program you wrote. Show instead how your career has progressed in scope and responsibility. Explain how your background and experiences have molded you into someone who is now uniquely qualified to fill the position being discussed.

If you are asked back for a second interview with the same decision maker, there's a good chance an offer will be forthcoming.

TAKE NOTES

This is a stressful time, and it's perfectly acceptable to take notes. You can use them to help you write intelligent follow-up letters and prepare for subsequent interviews. Have a clean, legal-size pad and a handsome pen in your briefcase for this purpose.

IF IN DOUBT, ASK

If you don't understand the question, politely ask the interviewer to rephrase it. After you've answered, don't be afraid to inquire whether your answer was satisfactory.

GET THE INTERVIEWER OUT FROM BEHIND THE DESK

This technique can lead, *so effectively,* into a successful interview that it deserves special attention. The desk is a barrier and an authority symbol. It comes between you and the interviewer and hinders your ability to establish rapport. By getting the interviewer out from behind this barrier, the interview becomes "you and me against the problem," instead of "you against me."

If there is a couch—or two comfortable chairs side by side—in the office, endeavor to sit with the interviewer. Perhaps you can lure the interviewer away from his perch by asking him to point out the various parts of the facility visible from his office window. Would it be possible to tour the area where you would be working? Could you meet the person who would be your immediate supervisor? Just how do they go about doing the job?

If the interviewer won't budge, perhaps you could walk around to his or her side of the desk to show samples of your work. Just don't move in too close. Everyone has a "comfort zone" that diminishes the longer they know someone, but you're not friends yet.

INTERVIEW THE INTERVIEWER

Too many job hunters beg for jobs, too few shop for them. Even if you're ready to take *any* job, don't betray this fact in your interview. Endeavor to determine if this position really would be a good match.

Before you accept any position, get the facts that will help you make an intelligent choice. This *could* be the last (and best) career decision you make, if you choose wisely.

Be discreet. Don't start out by asking what the salary and benefits are, and don't ask more questions than you answer. Working intelligent questions into your conversation, however, will give the interviewer a welcome break.

QUESTIONS TO ASK THE INTERVIEWER

ABOUT THE POSITION

To whom would I report?

What positions would report to me?

To whom does my immediate superior report?

What is the exact job description?

Is this description accurate?

What is the size of the department?

How is the department structured?

Where would I actually be working?

Why is this position available?

How much (if any) travel is involved?

Will I receive training? From whom? What does it consist of?

How long has the position been open?

How many have applied for the position?

Are any current company employees applying for this position?

What is the turnover rate for this position?

When will you make your decision? Who has the final say?

What is the company policy for performance reviews? When are they conducted? By whom?

How are raises and promotions decided?

What kind of growth and opportunity could I expect?

ABOUT THE COMPANY

What is the corporate philosophy?

Are there plans for expansion, relocation, or closings?

Are any of the workers unionized?

Are sales up, down, or the same as last year?

What new products or services is the company planning to introduce?

What new systems or procedures have been recently instituted?

What is the rate of employee turnover?

When did the present owner buy the company?

Is this a takeover or merger pending? If so, how will present management be affected?

If you've done your research thoroughly, you'll already know the answers to many of these questions. Don't interrogate. Choose the questions you most want answered and which seem appropriate to the interview situation. Don't persist with a line of questioning that appears to be annoying the interviewer. Some interviewers are not authorized to give the answers to many of the questions above.

As discussed in Chapter 2, "The New Economy," many over-50 job seekers are leaving large, public companies and seeking employment in smaller, younger, perhaps privately owned companies. A small, closely held corporation, particularly one where the founder is still involved in day-to-day management, is an environment far different from that of a corporate conglomerate.

If the owner is a strong presence in the company (most are), every idiosyncrasy he displays becomes company legend. Each pronouncement gets etched in stone. Day-to-day interactions with coworkers are governed less by written policies and more by something that can best be described as sibling rivalry and family "pecking order." Be sensitive to such an environment. Tailor your questions accordingly.

TYPES OF INTERVIEWS

There are four types of interviews. We'll discuss each one.

∎ *"Tell Me about Yourself"*—You'll recognize it immediately. The interviewer will ask this question within minutes of beginning the interview. She's giving you enough rope to hang yourself, so be careful.

Briefly review your background. Prepare a two-minute talk, outlining:

Where you were brought up

Your education

Your early work life

Your most relevant and recent experience.

Rehearse your talk, and time it.

Then, ask her what specifics she would like to know. Don't just babble. Pin her down and make her ask specific questions. Ask the questions you want answered. Turn your soliloquy into a productive conversation. If the interviewer never talks, she never has to commit herself to hiring you.

■ *"Listen to my story"*—This is the opposite of the "tell me about yourself" interview. It may take a while, but eventually you'll realize that the interviewer hasn't stopped talking since you entered the room. It can be extremely irritating to prepare yourself to make a dynamic presentation—and never get a word in edgewise.

You *must* get a word in edgewise. Interrupt politely: "Excuse me. I'm interested in that project. It sounds similar to a project I was involved in on my job. Let me tell you a little about what I did . . . " Keep the interviewer's attention on you.

■ *"Stress interview"*—There may be one interviewer, or several. Their purpose is to rapid fire questions at you and see how you handle it. Luckily, it doesn't occur that often.

Don't buckle. Stay in control. If it's going too fast and you feel yourself slipping, take a deep breath, pull yourself up, and take charge. Say, with good humor and a smile, "Whoa, Gentlemen (Ladies, Folks). Let's take it one question at a time. I'm here to give you an interview, not a confession. Now, Mr. Adams, you were asking about my MIS background. . ."

▮ *"Normal interview"*—No stress, beating around the bushes, or double-edged questions. It's a solid, informative session aimed at filling a position with the most desirable applicant.

Review the next section to prepare for this type of interview.

FIELDING THE QUESTIONS

The toughest questions you'll answer are those that deal with the issue of your age. They will be phrased many ways, but they'll all be asking the same thing: "Aren't you too old for this job?"

You cannot avoid this issue. It *will* come up, so prepare to deal with it—and dispense with it. You *can* convince the interviewer—any interviewer—that you're a contender. Here's my favorite version of the "age" question:

Aren't you overqualified for this position?

Here's *my* answer:

I'm glad you perceived my strong qualifications right away. I'm probably more qualified than other candidates for this job, but I don't believe I'm *over*qualified. Because of my skill level, I'll be able to take over the position with little or no 'down time.' Pretty soon, I'll be using my experience to bring more to the position than you've come to expect. In that way, I remain challenged and the company gets a better return on its investment.

If they persist in a similar line of questioning, you can face the issue head on and say:

I sense that you're concerned about my age. I assure you I'm in excellent health and in the prime of my career. I'm not ready to put all my knowledge and skills on the shelf for many years to come. There's a lot I still plan to do, and if this company hires me it will be getting the *benefit* of my experience and knowledge. That's an asset, *not* a liability.

Be sure to reinforce your statements by highlighting recent achievements that demonstrate how you've kept your skills and knowledge current. Don't be "yesterday's news."

Think about your answers, and write them down. Review your answers, rehearse them in front of a mirror, talk into a tape recorder. You've been forewarned, so be forearmed. There's no need to suffer a surprise attack and be beaten back.

THE APPLICANT'S VALUE TO THE COMPANY

- ∎ What qualities and skills do you believe the successful candidate should have for the position we have open? (They're trying to determine how accurate are your insights regarding the job requirements.)
- ∎ Why do you want to work for us? (Testing your commitment.)
- ∎ What can *you* bring to this job that others can't?
- ∎ Why should we hire you?
- ∎ How do you feel about having been let go by your last company? (They're probing your mental attitude and composure. Remember, stay *positive*. Practice this one till it's perfect.)
- ∎ Why haven't you found a position yet?
- ∎ Does anything about the job (industry, company) bother you?
- ∎ What interests you about us?

EXPERIENCE AND BACKGROUND

- ∎ Tell me about the positions you have held.
- ∎ What things in your work experience have you enjoyed the most? The least? (Highlight those that relate to the position being discussed.)
- ∎ What kinds of activities do you pursue (outside your job) to stay active in your profession? (Mention professional associations, seminars, courses, reading, research, articles contributed to journals, and so forth. Show that you keep current.)

‖ Tell me about problems you've solved on your job.

‖ What new skills have you acquired on the job?

‖ Tell me about the biggest success you've had at work.

PERSONAL/FAMILY

‖ What do you do for relaxation?

‖ What are your hobbies, interests, other outside activities? (Focus on your involvement in your community, leadership, etc.)

‖ How much work have you missed due to: illness? personal problems?

‖ What sports and recreational activities do you pursue? (Emphasize more rigorous, health-oriented pursuits: For example: golf over bridge, racquetball over reading.)

‖ What do your family members think of your making a job change now? (They're probing family relationships, determining if your role in the family is dominant. Give a balanced answer that doesn't play into any prejudices.)

‖ Are you willing to relocate?

‖ Are you able to travel?

‖ Can you work long hours?

GOALS AND MOTIVATION

‖ Why did you leave your last job? The one before that?

‖ How did you get the job with _____?

‖ What are your plans for the rest of your career? Where will you be five years from now?

‖ What are your current earnings? What do you expect to earn with us? (Don't give a figure for "expectations." It depends on the job requirements.)

‖ What is your definition of success?

▮▮ Are you satisfied with your career progress to date?

▮▮ Why haven't you advanced farther in your career?

▮▮ Are you considering any different positions?

▮▮ What type of work are you looking for?

WORK STYLE AND INTERPERSONAL SKILLS

▮▮ How do you plan and organize your work?

▮▮ Describe the best manager you have had.

▮▮ What would your current (former) supervisor say are your strengths and weaknesses.

▮▮ How would your subordinates describe you as a manager?

▮▮ What do you see as your strengths and weaknesses?

▮▮ Give five adjectives to describe yourself.

▮▮ What is the definition of a good manager?

▮▮ Have you ever fired anyone? Why? How did you go about it?

▮▮ If you had had your boss's job, what would you have done differently?

▮▮ Describe a subordinate of whom you are most proud. What did you have to do with his (her) success?

THE QUESTION OF MONEY

Who should bring up the subject of compensation, and when? The answer varies with each situation. You should have an idea about the salary range for the position before you interview. If it hasn't been advertised, make some discreet telephone inquiries. If the personnel office won't talk, perhaps someone on your network will. At any rate, you have enough career experience to know if you're interviewing for a job "in the ballpark" of your salary requirements.

Go into the interview knowing what the minimum you need is. Be realistic. What will it cost you to make this move? Will you lose benefits? If a relocation is involved, what kind of real estate market are you moving from—and to? If you're moving from Kansas City to Los Angeles, your cost of living may double. Do your research in advance, know what you need, but don't lay your cards on the table—yet.

If the interviewer doesn't bring up the subject, ask "What salary range has been assigned to this position?" (Realize, too, that salary ranges go out the window every day when the right candidate comes along.) "What nonmonetary forms of compensation (benefits, perks) typically accompany this job?"

Ask about timetable for performance evaluation and salary reviews. Depending on when the first raise is given, total first-year compensation could be a much higher figure than the opening salary. Don't spend the entire interview talking about money. Make your best presentation, so they're sold on you before they have to start loosening the purse strings.

CLOSING THE INTERVIEW

When the questions are asked and answered, the ball is back in your possession. It is now time for your star play: the closing. Make your last impression a lasting one.

1. Ask final questions pertaining to the job and its responsibilities. Clear up any confusion you have about the job requirements now.

2. Ask the interviewer if there is anything else he would like to ask, any more information you could provide in order to aid the hiring decision. This technique may reveal what the interviewer thinks about you at this point.

3. If you're interested, make it known. "I think we have a good match here, and I'm looking forward to exploring it further."

4. Make a summary statement that sells. You've got the floor. Make it pay. Lawyers put hours of work into their

brief closing statements. Rehearse your job-winning words.

5. Ask when a decision will be made. If the interviewer is specific, write it down. Say: "I'll call you then to follow up. I don't want to miss your call."

6. Open your parachute. If you have the distinct feeling that "disaster" is too mild an adjective for this interview, admit it. It can't get any worse, and taking a risk could save the day. Ask the interviewer where you went wrong. After all, you have other interviews and you want to improve your presentation because you have a lot to offer. Only the coldest heart will not warm at a sincere admission of failure and an honest effort to improve. Sometimes, this tactic will get the interviewer to re-evaluate you—even reinterview you. Even if it doesn't, the feedback will help you do better next time.

INTERVIEW FOLLOW-UP

Just as you organized your research notes on prospective companies for ready access and recorded your networking activities in a log, you must also track your interviewing activities and follow up. The final quarter of the game is the most important. Don't drop the ball when you're ahead.

At the end of every day, review your notes on each interview. Evaluate yourself honestly. What "grade" would you give yourself for poise, confidence, presence, the way you answered questions? What would you have done differently? Are there any questions you neglected to ask? List the names of the people you met (correct spelling, titles!) and jot down a few details about each while the interview is still fresh in your memory.

THE FOLLOW-UP LETTER

Write a follow-up letter to *everyone* you met during your interview—and send it within a day or two of the interview. He who hesitates is not hired.

Don't settle for the standard "thank you for your time," notes that anyone can send. Write a thoughtful letter that sets you apart, shows your insights into the company and the position, and lets them know you're the man (woman) for the job.

The letter should be no more than one page, typed, and in the same format as that given for cover letters earlier in this chapter. Here's your chance to sell yourself one more time. You may be one candidate among dozens, but your letter can help you be the one they like best. An example is shown in Figure 11-2.

Figure 11-2   Follow-Up Thank You Letter

Gary J. Allen
4599 Sinclair Street
Windsor, Ontario M6J X19

December 15, 1990

Robert Schultze, Comptroller
American Tool Company
213 W. 15th Street
Birmingham, MI 48009

Dear Mr. Schultze:

Thank you for interviewing me for the position of accounting supervisor at American Tool Company. I was impressed with your operation. It would be a pleasure to work in such a well run environment.

I neglected to mention that I have enrolled in a computer class beginning next month that will involve the newer JBM equipment you spoke about. Having been involved in several "change-overs" such as the one you anticipate, I am well aware of what needs to be done to make it go smoothly.

As we agreed, I will contact you Friday for your decision. Again, I appreciate your time and consideration.

Sincerely,

Gary Allen

THE FOLLOW-UP TELEPHONE CALL

It's not over yet. If they don't call you within a week of sending your follow-up letter, take a deep breath and call the interviewer, whether or not you said in your letter that you would call.

There could be many reasons for the delay. They could still be interviewing. Perhaps they're still evaluating all the candidates and trying to come to a decision. Let them know you are considering other positions, and you must make a decision shortly. Companies tend to be interested in the candidates other companies are interested in.

THE OFFER

Look before you leap! When offered a position, thank them sincerely and request a day or two to consider the offer. Several things can happen during this time. You could receive another offer. Even if you preferred the first job, you could use the existence of another offer as a negotiating wedge for compensation and other job conditions.

Any career move entails a great deal of work and stress. Is this offer worth everything that will be required of you? If relocation is involved, do the compensation and benefits really make it worth it?

Never start a new job at a disadvantage. Chances are, they'll never like you as much as they do at this minute. Use that fact to carve a niche for yourself. Get what you need to do the job well. This isn't a *gift,* it's an exchange.

Neither accept nor reject an offer right away. A day or two of reflection places things in a different light.

PUT IT IN WRITING

A written record of your offer and acceptance prevents misunderstandings and hard feelings down the line. A written offer protects you and gives you something tangible to evaluate.

Most companies do this as a matter of course, but if it isn't mentioned, simply say cheerfully, "This sounds great. When can you get that on paper and in the mail so that I can look at it and let you know?"

When you've decided, telephone the person responsible for hiring you and let him or her know. Whether you've accepted or declined, confirm it in writing. Your written acceptance letter may contain any verbal changes made to the offer.

Even if you've declined, always say "good-by" with style. The world of work is amazingly small and closely knit. You may meet these people again, and you want them to think well of you.

Good luck and good interviewing.

NOTE

1. Jeffrey G. Allen, J.D., C.P.C., *The Complete Q&A Job Interview Book,* New York: John Wiley, 1988, p. 2.

12 | From Military to Civilian Employment

When Johnny comes marching home again . . . the men will cheer and the boys will shout and the ladies they will all turn out . . .

Will that also be true for the job market? How will the work place receive today's over-50 career military man or woman making the transition to private sector employment? This chapter explores the unique situation of changing from a military to a corporate uniform after 50 and gives specific guidance for making the transition a smooth and successful one.

FACTORS IN YOUR FAVOR

Changing jobs after 50 can be difficult for anyone. But finishing up a career in the military service and taking up a new one on the "outside" requires exceptional stamina, initiative, and moxie. Luckily, you've got all three. There are three more factors in your favor:

1. The increased popularity of American service men and women at home

2. The country's transition from a manufacturing to a service economy
3. Special skills and experience gained from a military career that are highly prized in the private sector.

Let's explore these advantages in more detail.

INCREASED POPULARITY

The country's attitude toward its military personnel is perhaps more positive now than it has been at any time since the end of World War II. This may be due in part to better public relations campaigns by the armed services; it may be the result of a national guilt about the way returning Vietnam veterans were treated through most of the 1970s and 1980s. Whatever the reason, this new approval rating has some spillover into the work place. While you probably won't be greeted with a tickertape parade, you'll likely encounter less anti-military sentiment than would have been the case 15, 10, even 5 years ago.

SERVICE-BASED ECONOMY

Military work is service work—administering, supplying, repairing, supporting, procuring, accounting, and much more—on a large scale. Labor-intensive service industries are the fastest growing sector of the U.S. economy. Military work is nothing if not labor intensive and service-oriented. The only thing missing in the military picture is profit. It will be up to you to prove your (as yet untested) ability to mind the bottom line.

SPECIAL SKILLS

An over-50 job seeker whose career has been primarily a military one may possess greater leadership experience and ability, better discipline and physical conditioning, and—in some cases—a broader perspective than that of someone with an equivalent civilian career.

These special skills can give you a competitive edge in the job market if you know how to communicate and display them properly. If you don't put the right spin on your military skills, however, they could work against you. More on that in the next section.

PROBLEMS TO PONDER

For many, the military is not just a career, it is a way of life. And it is a way of life foreign to almost everyone who has not experienced it. If you are a high-ranking officer leaving after many years of service, you may have been exposed to both challenges and perks that have no parallel in the civilian world. Both could serve to insulate you and distance you from the other world you are about to enter.

It will help you in your search if you adjust for the fact that most people you will meet in your job search do not have a clue about the life you're leaving. If you confine your conversation to military anecdotes, and pepper your resume and speech with arcane acronyms and other jargon, your entire presentation will be, effectively, in a foreign language.

Further, the civilian employer rarely knows how to value your military experience and equate it to tasks and responsibilities in a civilian occupation. It is your job to carefully assess your achievements and translate them into recognizable accomplishments for your first civilian resume and interviews. A list of the resources available to help you do this appears later in the chapter.

Let's first take a look at how you can capitalize on the past 25 years to make the next 25 even more productive and rewarding.

MAKING A SMOOTH TRANSITION

Everything written in this book about the over-50 job candidate, and all the job-getting strategies prescribed, apply equally to the military-to-civilian job seeker. There are, however, *additional* strategies for your success. Begin to employ them *before* you drive out the base gate for the last time.

ADVANCE PLANNING

Despite the advice from folks who tell you, "Employers don't even want to talk to you until you're out," there are several important steps to take before you leave the service. You want your forces fully deployed before you begin battle. Studies have shown that the more time and effort invested in planning and preparing for a civilian career, the better the returns. A checklist for your planning is shown in Figure 12–1.

TAKING STOCK: YOUR SKILLS INVENTORY

In following the strategies in Chapter 5 for preparing a skills inventory, you will need a few additional resources to enable you to inventory your skills in civilian—rather than military—terms.

Two of these are the *Dictionary of Occupational Titles* (DOT) and the *Occupational Outlook Handbook,* both published by the U.S. Bureau of Labor Statistics. The DOT lists and describes more than 30,000 civilian jobs. The *Handbook* provides the outlook for and training required for each job. A companion publication, *Occupational Outlook Quarterly,* supplies similar information on a more current basis. All three can be found in most public libraries.

Other helpful publications can be found in the library, as well. They include *The Encyclopedia of Managerial Job Descriptions* and *The Encyclopedia of Prewritten Job Descriptions,* both published by Business Research Publications.

Finally, the basic reference used by the military is the *Military-Civilian Occupational Source Book,* DOD 1304, 12Y, which is held by most commands and recruiting centers. This can also help you when you arrive at the point of choosing a job.

TARGETING THE RIGHT CIVILIAN JOB

The DOD publication just mentioned gives occupational code numbers (OCNs) for every every military specialty. Caroline

Figure 12-1 Checklist for Preparing for a Civilian Career

☐ Get all your military records in order. Make good, clear photocopies of all commendations, training certifications, and personnel, medical, and dental records. Carefully review, before signing, your Form DD 214, Certificate of Discharge.

☐ Meet with a counselor at the base educational center to discuss how your military training can be converted to college credits. Have a transcript prepared. Discuss future educational options and benefits, and take whatever aptitude tests —Scholastic Aptitude Test (SAT), Graduate Record Exam (GRE), Legal Scholastic Aptitude Test (LSAT), and so forth— you may need.

☐ If you're still a year or two away from discharge or retirement and you're in a place where education is available to you, take as many courses as you can that apply to your chosen civilian career. If you're still undecided, take survey courses in several disciplines to help you narrow your choice.

☐ Read what people on the outside are reading. If you're planning to transfer your military leadership experience to a civilian managerial career, read management periodicals, current business books, and trade journals to "bone up" on your prospective career. Check Chapter 7 for resource lists.

☐ Prepare financially for your time in transition. Where will you live while looking for work? How will you finance your job search? Are your savings adequate to sustain you through a long job search? Statistically, in the job market of the 1990s you can anticipate a search time of one-and-one-half months for every $10,000 of salary sought.

☐ Investigate unemployment compensation benefits. Veterans are entitled to receive them. While some view an unemployment check as welfare, that is not what it is at all. It is a portion of what you earned coming back to you. State laws vary as to how a veteran collects unemployment compensation; some require you to collect in person while others mail your check. Consider these laws when choosing a state in which to reside while looking for work, particularly if there is a possibility that you will be moving again anyway.

and Richard De Prez, in their *Resume and Job Hunting Guide for Present and Future Veterans* write:

> Some civilian relationships are quite extensive. For example, Army Clinical Specialists, Air Force Aeromedical Specialists, and Coast Guard and Navy Hospital Corpsmen might find they are qualified in anywhere from one to twelve medical positions. The same holds true for storekeepers, radio operators, photographers, data processors, personnel specialists, and others. While the listing does not cover each matching civilian job that actually exists, it does indicate the range of occupational choices available to you.
>
> It is important to cover as many civilian options as possible in order to broaden your marketability. Try to identify all the OCNs which can relate legitimately to your military speciality, and select the one(s) which offer the best use of your experience and talents.[1]
>
> Some military specialties are easily converted to civilian jobs. For example:

Military Occupation	*Civilian Occupation*
Accounting Specialist	Bookkeeper, Accounting Clerk
Administrative Support Specialist	Clerk Typist, Secretary, Administrative Assistant
Legal Technician	Paralegal, Legal Assistant
Disbursement Officer	Payroll Supervisor

To help about-to-be-former military people select a job objective, DePrez and DePrez (in the guide just mentioned) define three broad employment groups. Decide which of the three best describes you:

1. *The specialists*—Those veterans with a military specialty who wish to pursue the civilian equivalent of that specialty.

2. *The generalists*—Those veterans seeking a total career change by moving into a civilian job situation bearing little or no resemblance to jobs performed while in military service.

3. *The career-changers*—Those veterans seeking a total career change by moving into a civilian job situation bearing little or no resemblance to jobs performed while in military service.[2]

Earlier I discussed how to target your objective if you're a specialist—by using the various references listed to "translate" your military specialty to a civilian one. According to De Prez and De Prez, a generalist is:

> [A]n experienced officer or enlisted person whose responsibilities have expanded his primary military specialty into areas of administration, management, and operations involving leadership and command. The more senior, the more "general." These are the men and women who wish to continue their management role in some civilian capacity.
>
> The likeliest candidates are Army and Marine Corps officers and enlisted personnel in combat arms, line officers in the Navy and Coast Guard, and officers with flight designations. E-7s, E-8s, and E-9s of all services in supervisory roles are legitimate candidates for this group.
>
> The generalist has a lot going for him. He is a proven performer who has survived stiff competition to arrive at positions of responsibility and trust. He has become adept at managing people—a skill that commends itself strongly to a complex world which places increasing value on the importance of interpersonal relationships.[3]

Examine the preceding passage carefully. It contains some key words and phrases for your resumes and correspondence as well as your interview presentation.

The over-50 career-changer faces the toughest challenge. This job-seeker will have to work especially hard to start a new career. It will help if you can demonstrate effectiveness and the "proven ability to transfer skills" during your military years.

For example, in applying for a civilian position as general manager of a sports stadium, one former marine cited his "leadership and organizational ability under diverse and often difficult circumstances,"[4] as well as his after-work pursuits of organizing base recreational programs, as qualifications for an entirely new kind of job. Such various other duties as troop leader, manager of buildings, grounds, and military material, and responsibility for the security of nuclear weapons on board two aircraft carriers, all helped this career-changer achieve his objective. He got the job.

Career changers may find they have to take a position similar to what they did in the service, temporarily, with the aim of transferring to a different occupation within the employer's organization. That is what happened in the following case:

> John W., 51, retired from the Navy with the rank of Master Chief Petty Officer after 30 years of service aboard Naval nuclear submarines as a nuclear reactor operator. Desiring a change of scenery, John at first avoided applying to commercial nuclear power companies. Throughout his Navy years he had studied alternative energy and environmental energy issues, and that is the direction in which he targeted his search.
>
> Interviews and offers were almost nonexistent, however, except in commercial nuclear power plants. So great was the need for his skills and experience, he didn't even have to apply. The companies sought *him* out. Finally, he accepted a position in reactor operations with a large utility company, but he didn't give up hope.
>
> John read every job posting and worked with the company's career counselors; and, within the first year, he had transferred to the company's energy management services division as an energy consultant. Now his years of independent study are paying off, as he works on programs to help the utility's customers conserve energy.

John's experience shows that, sometimes, all it takes is a foot in the door—any door—to eventually find the right opportunity.

HOME IS WHERE THE JOB IS?

For some military-to-civilians, job location takes precedence over the job itself. According to De Prez and De Prez,

> Demographic studies show that retirees are especially prone to colonize in choice geographic locations distinguished by climate, leisure opportunities, and access to military facilities.
>
> For the generalist in this group, the problem is significant. Military density increases the competition for management positions, and his competitors are veterans with experience and skills similar to his own. For the generalist "willing to relocate," the prospects are brighter.[5]

Review Chapter 6 for help in deciding just exactly where you'll call "home" from now on.

RESEARCHING YOUR SEARCH

All the strategies in Chapter 7 for researching a target employer apply to you, the military-to-civilian job changer. There is another form of research that is especially helpful to someone who hasn't been inside the civilian work force for a number of years. It involves getting inside and taking a look around to familiarize yourself with the way business is done.

De Prez and De Prez advise veterans and future veterans to be aware of the differences between the way business is conducted in the military and civilian worlds. They claim there are "[t]wo noteworthy distinctions: (1) the preference in civilian organizations for the participative management style, and (2) the universal existence of the profit motive in the private sector."[6]

If you're a military generalist seeking to transfer your leadership experience to a civilian management career, it would benefit you to read up on the current trend in American business to the "matrix" organization and the "flattening of the pyramid," where management levels have been eliminated and objectives are ostensibly accomplished through cooperation and teamwork rather than the giving and following of orders—as is done in older, hierarchical types of organization. A good book along those lines is *Influence Without Authority* by Allan Cohen and David Bradford (New York: John Wiley, 1984).

Use any contacts you have, as well, to get an inside look at private sector employers. You can capitalize on the pro-military sentiment mentioned earlier to accomplish your in-person research. If you've been involved with defense contractors and suppliers, contact officials in those companies to get the names of managers in other, non-defense-related industries. (Defense contractors tend to operate too much like the military to give you any insights into the private sector.) Use whatever "in" you have to get inside target companies, and note not just how the business operates, but also how colleagues act, interact, talk, dress, and solve problems. You will gather important ammunition to use in preparing your resume and giving interviews. You may even pick up several good job leads.

A NOTE ABOUT DISCRIMINATION

As with age discrimination (discussed in Chapter 8), anti-military bias is not as blatant as in the past, but it still exists. Some good advice on how to defeat this prejudice from Caroline and Richard De Prez:

> Take off your uniform . . . figuratively as well as literally. And make every effort to convince civilian employers that the action was complete and final. Be careful of the image you project in speech, manners, appearance, and in the language of the resume. Employers need to know you're not there to throw your weight around, nor are you looking for a soft berth with minimal demands. Instead, you have a valid contribution to make and the qualifications to back it up.[7]

YOUR IMAGE AND ATTITUDE

This last issue brings us to the job search itself. As an "outsider" attempting to enter a new world, you will need to be especially aware of the image and attitude you project. While it may be relatively easy to learn to prepare your resume and correspondence in accepted civilian business formats rather than in the military style you have grown accustomed to, it may be more difficult to change 30 years of interpersonal behavior. You may need the guidance and advice of trusted friends and family members to help you change the habits of military life without lessening your well earned pride in your military accomplishments.

Tell those around you to comment when you lapse into military talk. All those initials and "0900 hours" will perplex most interviewers. If you're in the habit of giving orders and expecting them to be obeyed, work on consensus-building instead. The development of this skill can start with something as simple as what the family will have for dinner or do over the weekend.

To help get you ready for those first interviews and civilian career encounters, begin to move away from the comfort of your accustomed military existence. If you're still living on base in preparation for separation, spend more time shopping and recreating off base. (You'll need to get adjusted to the prices, anyway.) Go out and purchase your interview suit, and wear it out and about. To help you see why this is important, try to picture a career business executive putting on a colonel's uniform, complete with ribbons, for the first time and wearing *it* to work. Whether he realized it or not, he'd probably look pretty awkward.

Give your interview presentations in front of the mirror. Use videotaping, role-playing, or whatever techniques work best for you so that you can objectively evaluate how you come across to others. If you can't be objective, find someone to work with you who can be.

The Retired Officers Association (TROA) offers an excellent booklet for former and soon-to-be-former service people. It is entitled "Marketing Yourself for a Second Career" and is packed with helpful information as well as some good samples of "civilianized" resumes. Send $3.00 with your request to

The Retired Officers Association (TROA)
201 North Washington Street
Alexandria, VA 22314-2529
(703) 549-2311

If you're still in the service, you can receive the booklet at no charge by attending one of TROA's free lectures on the same topic. Your base Public Information Officer will have the date and place of the next scheduled lecture.

In addition, for a $20 annual membership fee, the TROA Officer Placement Service (TOPS) offers a monthy career newsletter, resume preparation assistance, career counseling, access to TROA's Career Research Center, and job referral information.

YOUR CIVILIAN RESUME AND CORRESPONDENCE

Follow all the guidelines in Chapter 9 for preparing resumes, cover letters, and thank you letters. Spell out titles and place names: U.S. Army, Europe, not USAREUR. November 21, 1990, not 21 NOV 90. Use an accepted *civilian* business format for all correspondence.

Do not include your social security number, nor a lengthy list of medals and commendations. Mention any special commendations that are relevant to your current job objective in the employment history section of the resume. Be sure to mention fluency in any foreign languages and other special skills that will be useful to your prospective employer.

You must take special care with your resume. Do not rely on a resume-writing service. No one but you can do it right, but it will help if someone with extensive civilian work place experience can critique the completed resume.

INTERVIEWING

As I do with all job seekers, I recommend you schedule a few "throwaway" interviews at first—for jobs that you really don't want. This real-life practice helps you to polish your interview style without causing depression if you don't immediately get an offer. Be aggressive in scheduling interviews right away. When one of those first, not-so-important interviews is over, say candidly, "How did I do? I would appreciate your assessment of my interview style."

Remember my earlier caution about your bearing, appearance, and attitude. Work on eliminating familiar phrases from your speech. Your appointment is at 9:00 A.M., not "0900 hours." Drop the "sir" and "ma'am," too. Your next boss will be known as "Bill" or "Ruth" even to the workers on the shop floor. At the same time, relax the posture and project yourself as friendly, approachable, and interesting.

Follow all the guidelines for the interview uniform and accessories. Even if your black dress uniform shoes are still in good shape, visit a shoe store and buy an up-to-date pair of civilian shoes. Men, let your hair grow a bit and then visit a barber or hairstylist on the outside who can give you a neat, businesslike look.

OTHER RESOURCES

Finally, here is a list of organizations that provide career counseling (and, in some cases, resume services, job fairs, and job banks) to military personnel, retirees, and dischargees:

❚ American Legion
National Economic Commission
1608 K Street NW
Washington, DC 20006

❚ Air Force Sergeants Association
4211 Auth Road
Suitland, MD 20746

❚ Association of the United States Army
2425 Wilson Blvd.
Arlington, VA 22201

❚ Disabled American Veterans
3725 Alexander Road
Cold Spring, KY 41076

❚ Fleet Service Association
1303 Hampshire Ave. NW
Washington, DC 20036

❚ Marine Corps Reserve Officers Association
201 North Washington Street
Alexandria, VA 22314

❚ Naval Reserve Association
1619 King Street
Alexandria, VA 22314

❚ The Retired Officers Association
201 N. Washington St.
Arlington, VA 22314

❚ Veterans of Foreign Wars of the United States
200 Maryland Ave., NE
Washington, DC 20002

On behalf of your fellow citizens, I'd like to say "welcome home" and "thank you" for your lifetime of service to all of us.

As a fellow 50-plus worker, I wish you luck and satisfaction in your new career. May your new civilian uniform be a comfortable fit.

NOTES

1. De Prez, Caroline Steed, and De Prez, Richard J., USN (Ret.), *Resume and Job-Hunting Guide for Present & Future Veterans* (New York: Arco, 1984), pp. 18–19.

2. Ibid., p. 17.

3. Ibid., p. 19.

4. Ibid., p. 92.

5. Ibid., p. 20.

6. Ibid., p. 21.

7. Ibid., p. 21.

Conclusion

As work progressed on this book, I found myself thinking back more and more over my own career, and especially my own experiences with job hunting after 50. I share them with you in the hopes they will be a source of encouragement.

I made two job changes after 50. The first time I "changed" into a job similar to the one I had just left. I went from being vice-president of personnel for a small supermarket chain to director of personnel for a large drugstore chain. Both were excellent positions, challenging and rewarding financially as well as emotionally.

But nothing in my career history compares to the satisfaction I gained when I redirected my focus and joined The Transition Team. In "starting over" at about the age most people are "wrapping it up," I got a whole new lease on life. There's nothing like learning new skills, meeting new challenges, and making new friends to recharge your batteries. It's a veritable vocational fountain of youth.

My promotion from vice-president to president of The Transition Team came during work on this book. I like to think my experience is proof there are no plateaus, only higher peaks we refuse to see.

Won't you join me up here? The view is just grand.

Bibliography

ADEA Guidebook, copyright 1987 by the American Association of Retired Persons, Worker Equity Department, 1909 K. St. NW, Washington, DC 20049.

Allen, Jeffrey G., J.D., C.P.C., *The Complete Q&A Job Interview Book,* New York: John Wiley, 1988.

Allen, Jeffrey G., J.D., C.P.C., *Jeff Allen's Best: Get the Interview,* New York: John Wiley, 1990.

Allen, Jeffrey G., J.D., C.P.C., *Jeff Allen's Best: Win the Job,* New York: John Wiley, 1990.

Allen, Jeffrey G. J.D., C.P.C., *Surviving Corporate Downsizing,* New York: John Wiley, 1990.

Birsner, E. Patricia, *The 40+ Job Hunting Guide,* New York: Simon & Schuster, 1987.

De Prez, Caroline Steed and Richard J. De Prez, USN (Ret.), *Resume and Job-Hunting Guide for Present & Future Veterans* (New York: Arco, 1984).

"Head-Renting. How Hot New Niche Affects Search & Consulting," *Special Report* July 15/August 15, 1990, Copyright 1990 by *Executive Recruiter News,* Kennedy Publications, Fitzwilliam, NH.

Kennedy's Career Strategist, Vol. V., No. 2, February 1990.

Lewis, Tamar, "Too Much Retirement Time? A Move Is Afoot to Change It," *New York Times,* April 22, 1990.

Naisbitt, John, *Megatrends,* New York: Warner Books, 1982.

Pifer, Alan and Lydia Bronte, Eds., *Our Aging Society,* New York: W.W. Norton, 1986.

Professional Guide to Career Life Planning, Copyright 1986 by The Transition Team Network, Inc., The Transition Team, 3155 W. Big Beaver Road, Suite 117A, Troy, MI 48084.

Ryder, Gail A. *The Transition Team's Guide to Relocation,* Copyright 1986 The Transition Team Network, Inc., The Transition Team, 3155 W. Big Beaver Rd., Suite 117A, Troy, MI 48084.

Solomon, Jolie, "Drucker Puts Name on Nonprofit Group," *The Wall Street Journal,* Dow Jones, August 6, 1990.

Sonnenfeld, Jeffrey, "Dealing with the Aging Work Force," *Harvard Business Review,* November-December 1978.

"Tactics That Win Good Part-Time Jobs," *Changing Times,* May 1990, Vol. 44, No. 5.

Using the Experience of a Lifetime, American Association of Retired Persons, 1988. (AARP, 1909 K. Street NW, Washington, DC 20049)

Index